Four

Wo

GCSE

Geography

for WJEC A

Angela Calvert and Lubna Raza

HODDER EDUCATION
AN HACHETTE UK COMPANY

The Publishers would like to thank the following for permission to reproduce copyright material:

Photo credits

p. 3, 5, 52 *all* © Andy Owen; **p.72** © blickwinkel/Alamy

Every effort has been made to trace all copyright holders, but if any have been inadvertently overlooked the Publishers will be pleased to make the necessary arrangements at the first opportunity.

Hachette UK's policy is to use papers that are natural, renewable and recyclable products and made from wood grown in sustainable forests. The logging and manufacturing processes are expected to conform to the environmental regulations of the country of origin.

Orders: please contact Bookpoint Ltd, 130 Milton Park, Abingdon, Oxon OX14 4SB. Telephone: (44) 01235 827720. Fax: (44) 01235 400454. Lines are open 9.00–5.00, Monday to Saturday, with a 24-hour message answering service. Visit our website at www.hoddereducation.co.uk

First published in 2014 by
Hodder Education,
An Hachette UK company
338 Euston Road
London NW1 3BH

Impression number	5 4 3 2 1
Year	2018 2017 2016 2015 2014

Cover photo: Volcanic crater, New Hebrides, Australasia © Larry Dale Gordon/zefa/Corbis

Illustrations by Integra Software Services Pvt. Ltd

Typeset in 12 on 14pt ITC Stone Sans Medium by Integra Software Services Pvt. Ltd., Pondicherry, India

Printed by Hobbs the Printers, Totton, Hants

A catalogue record for this title is available from the British Library

ISBN: 978 14718 35 339

Contents

How to use this book

Welcome to this Geography workbook! It has been designed to be used alongside the *GCSE Geography for WJEC A Core* and *GCSE Geography for WJEC A Option Topics* textbooks. You will need a copy of one or both of these books open in front of you every time you use it. The books are full of maps, graphs, photographs and diagrams. This workbook bases its activities around these images.

The workbook is arranged in double pages. Each double page aims to tackle one or two geographical ideas, concepts or case studies. Sometimes the aim is to improve a geographical skill, such as using a map to describe a location. These aims are given as bullet points at the top of the left-hand page in each spread.

There is another important piece of information that you will need before you can start any activity in this workbook. You will need to study the pages in the *Core* or *Option Topics* book. These page numbers are given in the top right of each left-hand page of the workbook.

You can write your answer to a lot of the activities in this workbook on the pages themselves. There are also some spaces to record other kinds of answers such as labelling sketches or drawing graphs. In this way you will gradually build up a set of notes and drawings that cover a large part of your GCSE Geography course. However, there isn't space in this book for all of your notes! You will need to keep a notebook too. This will be useful for your own notes. You will certainly need to use your own notebook for longer pieces of writing. The activities where you need to record an answer in your own notebook are signposted with a symbol of a pencil ✎.

It's good to keep lists and to tick off targets as you achieve them. We have, therefore, included a learning review at the end of each double page. These boxes quite often include simple steps but do tick them off. Add them all together and you will see how much progress you can make over the two years you study GCSE Geography.

We hope you will find the workbook a useful way to organise your notes and test your understanding as you work your way through this GCSE Geography course.

Water

River processes

See Core Book pages 2–3

In these activities you will:

- show you know how rivers can shape the land
- practise the technique of labelling and annotating a photograph.

1 Rivers shape the landscape. They erode soil and rocks from one place. Rivers then transport this sediment along their course before depositing it to make a new landform.

a) Study Figure 2 on pages 2–3.

b) Identify the main features of the river at labels **A** and **C** and complete the third column in Figure A below.

c) Suggest how much energy the river has at **B** and **C** (on a scale of very high to very low) and complete the fourth column in Figure A.

d) Work in pairs to suggest what processes may be happening at **A, B** and **C** (there may be more than one process at each location). Choose from processes in the box:

deposition saltation hydraulic action suspension abrasion attrition slumping

	Part of the river channel	River features	Energy levels	Possible processes
A	Outer bend		Very high	
B	Middle	Riffle		
C	Inner bend			

Figure A Features and processes in the river shown in Figure 2 on pages 2–3

e) Use what you have learned to complete this conclusion:

At A the energy level of the river is .... The river is able to use this energy to erode the river cliff using the processes of ....

At C the energy level of the river is .... This means that the main processes here are ....

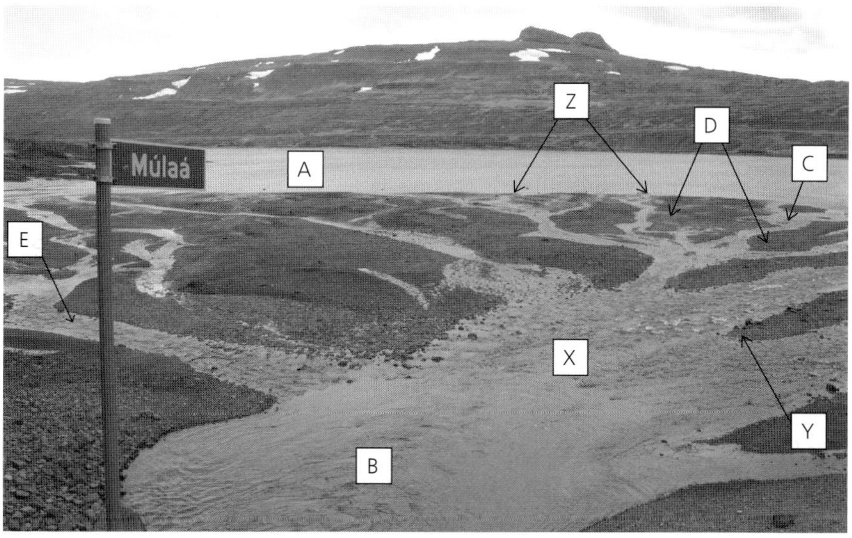

Figure B The processes at work in a delta

2 Study Figure 5 on page 3 and Figure B above. It shows a landform called a **delta**.

a) Labels are simple descriptive statements. Match the features below to points A, B, C, D and E by completing the table below.

Point	Label	Point	Label
	Main river channel		Mouth of the river
	Distributary channel		Lake
	Sand/gravel bar		

✎ **b)** To distribute something means to share it out. Using Figure B to help you, write your own definition for a distributary channel of a delta.

c) Add labels for two more distributary channels on Figure B.

3 Annotations are more complex statements than labels. For example, an annotation could describe the processes of erosion, transport and deposition responsible for creating different features of the landform in a photo.

a) Match the following annotations to points X, Y and Z by completing the table below.

Point	Annotation
	Water in the river channel transports sands by saltation and gravel by traction.
	Water velocity in the river channel slows down as it enters the lake. The flow of water is no longer strong enough to carry the sediment which is deposited.
	As the delta gets bigger sediment is deposited in the river channel blocking it. Then, during flood events, the river cuts a new distributary channel through the loose sands and gravels.

b) Use a coloured pen to highlight words that describe processes in the annotations.

✎ **c)** Suggest how the shape of the delta in Figure B might change over 20–50 years.

Review what you have learned

I can:

- label and annotate a photograph of a river landform. ☐

How are waterfalls formed?

See Core Book page 6

In these activities you will:

- practise using key words that describe river processes and river landforms
- show you know how waterfalls are formed and how they change over time.

1 Study the words in the box below. Some are river features; others are terms that describe the process of erosion or the transport of sediment in a river. Sort the words into the three categories. One of each has been done for you.

abrasion saltation river cliff traction delta suspension plunge pool attrition solution hydraulic action slip-off slope ox-bow lake corrosion meander

River features	Processes of erosion	Transportation
delta	*corrosion*	*saltation*

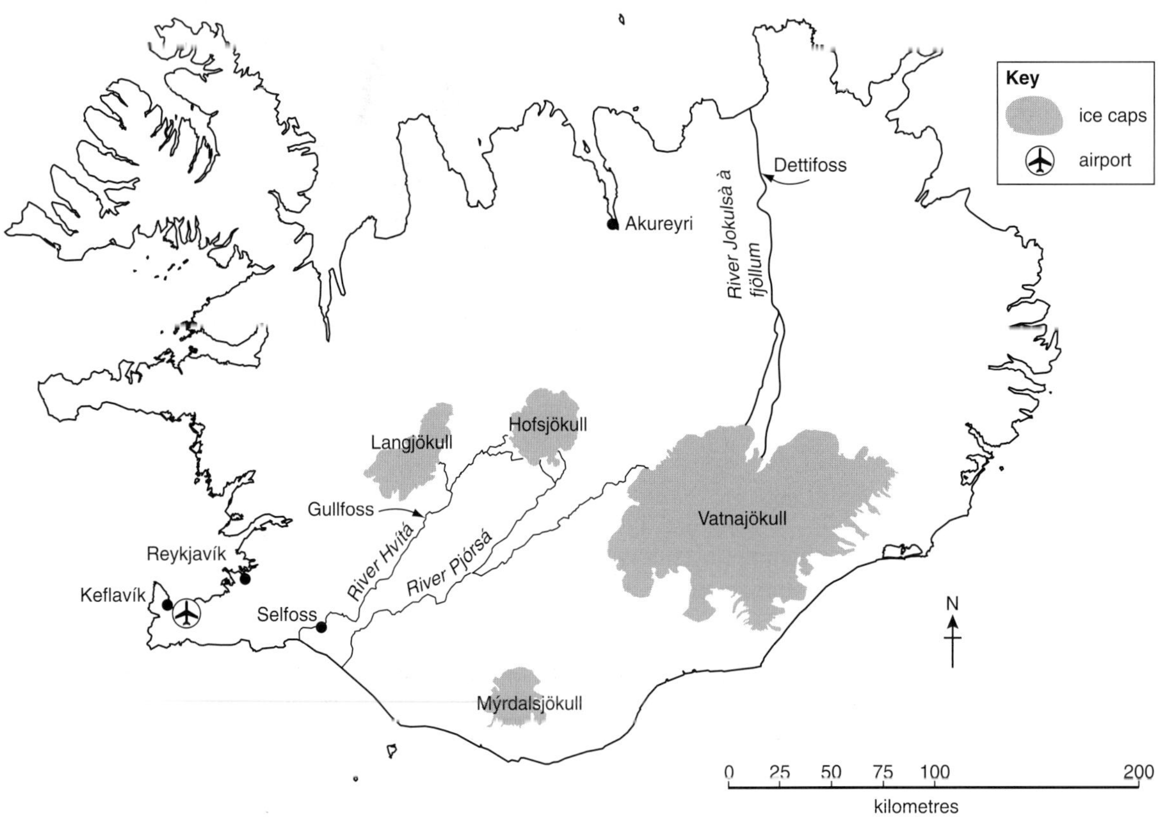

Figure A The location of Gullfoss, Iceland

2 Study Figure A opposite. Describe the location of Gullfoss: underline the correct words or complete the gaps in the passage below.

Gullfoss is a waterfall on the River **Hvítá** / **Pjórsá**. The source of this river is the ice cap known as **Myrdalsjökull** / **Hofsjökull** / **Vatnajökull**. The river's mouth is on the **east** / **north** / **south** coast. Gullfoss is .... km from Keflavik and .... km from Selfoss.

3 Figure B shows another large waterfall and gorge in Iceland close to Dettifoss.

a) Add the labels, below, to suitable places on Figure B.

1 A layer of harder rock that is difficult to erode	**2** Plunge pool where hydraulic action and abrasion are eroding the bed of the river
3 Overhang that could collapse in the future	**4** Steep sides of the gorge formed as the waterfall retreats

Figure B Waterfall at Dettifoss, Iceland

✎ **b)** Describe the location of Dettifoss.

✎ **4** Now draw your own series of diagrams to show how this waterfall will change over time. Add labels that explain what is happening. Use your own words as well as those from the box below.

erosion	**collapse**	**plunge pool**	**retreat**	**hydraulic action**	**gorge**

Review what you have learned

I can:

- use key words to do with river processes with confidence ☐
- describe the location of a place on a map using a scale line and compass ☐
- describe how river processes can change one landform over time. ☐

Flow of water through the drainage basin

See Core Book pages 9–11

In these activities you will:

- improve your knowledge of key words to do with drainage basins
- show that you know how water flows through different drainage basins.

1 Match the following key words to their definitions in the table below.

a) Precipitation
b) Transpiration
c) Evaporation
d) Condensation
e) Infiltration
f) Percolation
g) Overland flow
h) Throughflow
i) Groundwater flow
j) Interception

Key word	Definition
	Water moving through the ground below the water table
	When leaves prevent water from falling directly to the ground
	The movement of water from the ground surface into the soil
	The process where water changes state from gas to liquid
	The movement of water from the atmosphere to the ground
	The downhill flow of water through soil
	Water loss from plants through pores in the leaves
	The flow of water across the ground surface
	Where water changes state from liquid into vapour
	Movement of water out of the soil and into the rocks below

2 a) Add the key words a)–j) from activity 1 above to Figure A on page 7 to show the stores and flows in a natural drainage basin.

b) Draw arrows between the key terms to show how the water moves around the drainage basin.

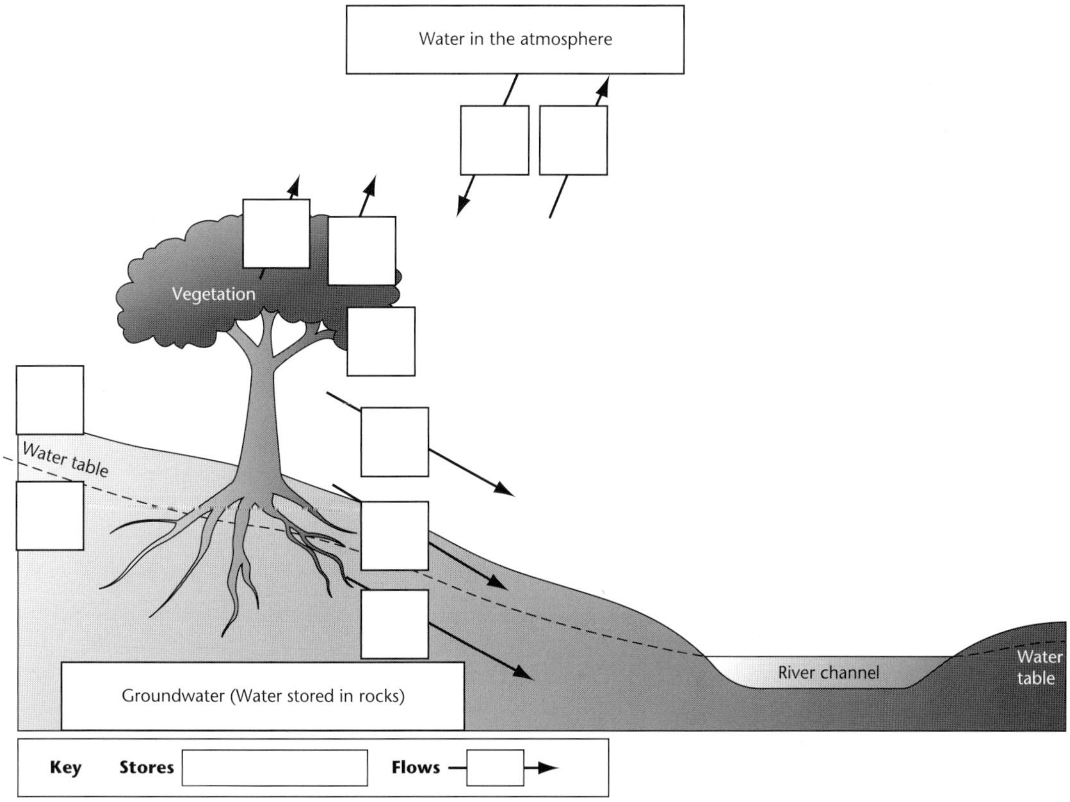

Figure A Movement of water through the drainage basin

3 Work in pairs for this activity. Study Figure 19 on page 10 and Figures 20 and 21 on page 11. Next, study Figure B below. Predict which hydrograph each catchment area would produce by writing A or B in the box next to each statement.

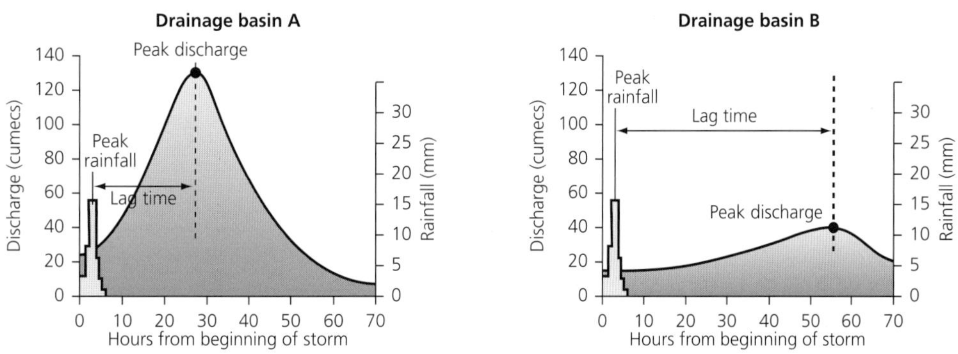

Figure B Hydrographs for two different catchment areas

Trees have been removed from this catchment area		Ground is impermeable and so water runs over the surface	
Catchment area with a forest		There are lots of roads and buildings in the catchment area	
Ground is permeable (water can soak into the ground)		The ground in the catchment area is frozen	

Review what you have learned

I can:

- describe how water moves through the drainage basin. ☐

Causes of floods

See Core Book pages 12–13, 17

In these activities you will:

- investigate why rivers flood
- analyse and compare two hydrographs.

1 Match the following key terms to features on the map in Figure A below. Write the correct letter next to the feature.

1. Catchment area = ...
2. Confluence = ...
3. Source = ...
4. Tributary = ...
5. Watershed = ...
6. Mouth = ...

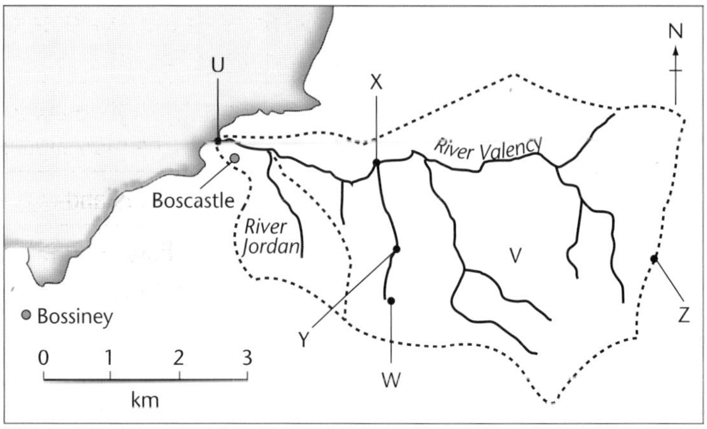

Figure A The drainage basin of the River Valency

2 The water level in the River Valency rose very quickly during the flood of 2004.

a) Study Figure 22 and Figure 23 on page 12.

b) Use the information in the orange boxes in Figure 23 to complete the right-hand column in Figure B below.

Reasons for the sudden rise in water level of the River Valency	Effect on the flow of water in the area (choose from the orange boxes in Figure 23)
1 Narrow valleys so ...	
2 Small areas of woodland so ...	
3 Steep slopes so ...	
4 Impermeable underlying rock so ...	
5 Upland areas are very close to Boscastle so ...	

Figure B Features of the drainage basin of the River Valency

c) Look at Figure 22 on page 12 carefully. Match each four-figure grid reference from the box below to the most appropriate description that follows.

1091	**1188**	**1190**	**1289**	**1291**

	Grid reference	Description
A		The flat summit of a hill with a height of 274m.
B		The contours are very close together showing that these south facing slopes are very steep.
C		Two of the Valency's tributaries meet in a confluence. The number of streams in this part of the map suggest that the geology is impermeable.
D		Small area of woodland of deciduous trees close to the river.
E		This tributary flows northwards towards the Valency. The contours make a sharp V shape indicating that there is no flood plain. This is a very narrow V-shaped valley.

3 Some floods are caused by melting snow, or long periods of rainfall, that saturate the ground with water. Study Figure 32 on page 17. Complete the sentences below using words from the box.

anticyclones	**40**	**depressions**	**snow**	**400**	**high**	**6th**
low	**rising**	**sinking**	**rain**	**120**	**3rd**	**180**	**165**

From 7th to 13th January there was a series of .... over Wales and the Midlands which led to wet weather. This type of weather event happens when the air has .... pressure and is .... and cooling which leads to clouds and rain. On 15th January, the rainfall was .... mm and the discharge of the River Severn was .... cumecs. By 17th January the discharge had increased to .... cumecs. One reason for the delay between the highest rainfall and the highest discharge is that the precipitation fell as .... and so took a few days to melt and flow to the river. The highest rainfall in this period happened on .... February at .... mm. The highest discharge of the River Severn in February was on .... February at .... cumecs.

Review what you have learned

I can:

- find evidence on an Ordnance Survey map that helps to explain flooding ☐
- read rainfall and discharge values from a hydrograph. ☐

Management of floods

See Core Book pages 14–21

In these activities you will:

- show you know how rivers can be managed to reduce flooding
- comment on the advantages and disadvantages of each management strategy.

1 The River Valency has been managed to reduce the risk of future floods. Study the features shown in Figure 26 on pages 14–15.

a) Explain how each of the features may reduce the risk of future floods. Complete the second column of the table below.

b) Label the hard engineering strategies in the table with H, and the soft engineering strategies with S. One has been done for you.

Management strategy	How it will reduce the risk of future floods	H or S?
Catchment management upstream, e.g. trimming back and removing trees from water's edge		S
Make the river channel wider and shallower just before the river enters the village		
Introduce a 'trash screen' which will stop debris like boulders, stones, trees and branches being washed into the culvert drain and blocking it		
New flood defence wall		
Widen river channel		
Lower (dig out) the river bed		
Remove lower bridge		

2 Flood management strategies have advantages and disadvantages.

a) Use the information on pages 14–20 to complete the table below to show the advantages and disadvantages of flood management strategies.

Flood management strategy	Advantages	Disadvantages
Creating natural sections of river that will encourage deposition upstream of towns that are at risk of flooding, e.g. River Valency, Boscastle, page 14		
Straightening river channel, e.g. River Valency, Boscastle, page 14		
Having a flood plan for residents, page 16		
Having demountable flood barriers, e.g. River Severn, Bewdley, page 17		
Not allowing building of more homes on the flood plain, page 18		
Blocking field drains and planting trees in the upland part of the drainage basin, e.g. River Severn, pages 20–21		

Review what you have learned

I can:

- give three facts about management of the River Valency at Boscastle ☐
- describe four different ways that rivers can be managed to reduce flooding ☐
- give one advantage and one disadvantage of one of these management strategies. ☐

Theme 2

Climate Change

What is the greenhouse effect?

See Core Book pages 24–25

In these activities you will:

- describe the stores and flows of the carbon cycle
- show you know the difference between causes and effects of climate change.

1 Carbon dioxide (CO_2) is a greenhouse gas. In order to understand the greenhouse effect we need to understand the carbon cycle. Study Figure A and the labels below the diagram.

a) Sort the labels into two types:

- carbon stores
- flows between stores

Colour the stores and flows labels using two contrasting colours and complete the key.

b) Add the number of each label to the correct place on Figure A below.

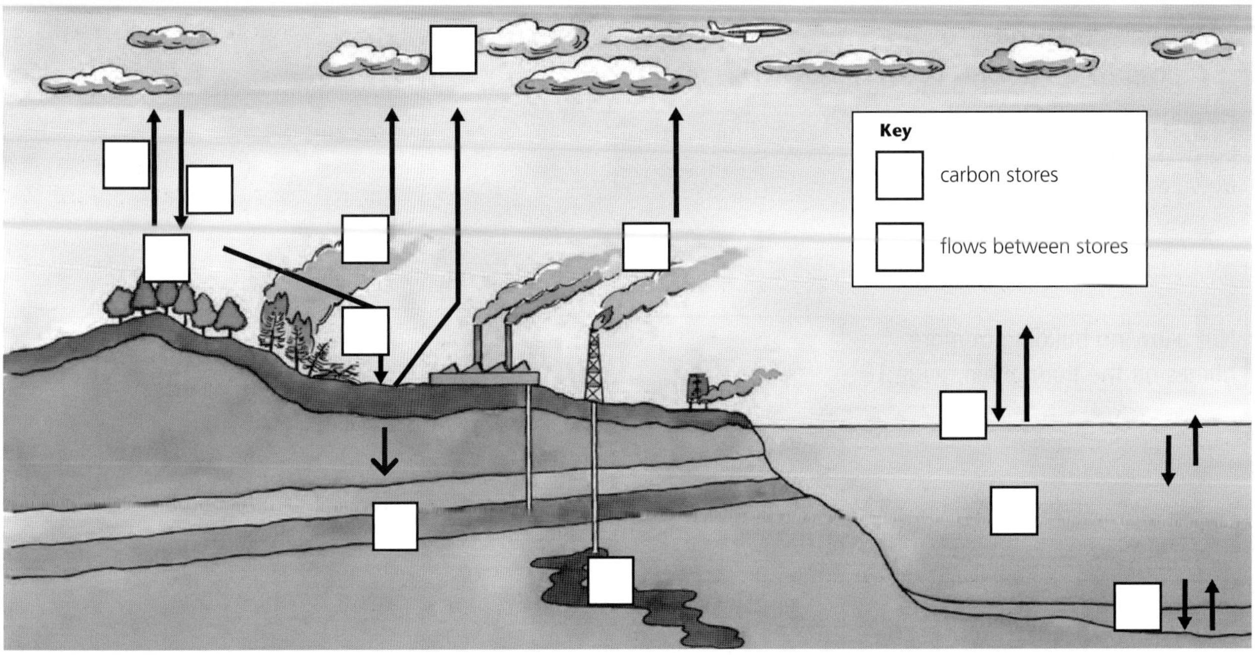

Figure A The carbon cycle

1. Fossil fuels, such as coal, gas and oil, store carbon for very long periods of time	**2.** Burning fossil fuels releases CO_2
3. Plants take in CO_2 during photosynthesis	**4.** Trees store carbon for hundreds of years
5. Decomposition of leaves / branches returns carbon to the soil	**6.** Living things breathe out CO_2 (respiration)
7. Burning trees releases CO_2	**8.** CO_2 dissolved in the ocean
9. CO_2 in the atmosphere	**10.** CO_2 moves between the ocean and atmosphere
11. Dead plant matter is compressed and turns to rock (lithification)	**12.** Dead algae and seashells contain carbon

2 Show you understand the carbon cycle.

a) Study each of the following lists of words. In each list, circle the odd one out.

					Common theme
1.	Fossil fuels	Forests	Atmosphere	Respiration	
2.	Cars	Airplanes	Wind turbines	Coal-fired power station	
3.	Photosynthesis	Respiration	Coal	Decomposition	
4.	Wind energy	Coal	Oil	Gas	
5.	Oxygen	Carbon dioxide	Methane	Water vapour	

b) For each list, state what the three remaining words have in common. Write your answer in the final column of the table.

3 Increasing the amount of carbon dioxide in the atmosphere seems to make the greenhouse effect stronger. This may be one explanation for recent climate change.

Read each statement in the table below. Decide whether it is a potential cause or effect of climate change. One has been done for you.

	Cause (C) or Effect (E)
Heating our homes by burning gas or coal	C
Burning coal in a power station	
Unusual weather	
Chopping down large areas of tropical rainforest	
More floods of coastlines	
Generating electricity by burning fossil fuels such as gas	

	Cause (C) or Effect (E)
Increasing numbers of flights	
Sea levels rising	
Ice melting in Greenland	
Burning coal in a factory to make steel	
More hurricanes and storms	
Increasing numbers of cars on the roads	

4 Study Figures 3 and 4 on page 24. Describe two ways that cutting down or burning large areas of tropical rainforest will affect the carbon cycle.

Review what you have learned

I can:

- describe flows and stores in the carbon cycle ☐
- give two causes and three effects of climate change. ☐

How conclusive is the evidence for climate change?

See Core Book pages 26–29

In these activities you will:

- examine the evidence for climate change
- interpret a climate graph
- review evidence from recent weather observations.

1 The majority of climate scientists agree that people's actions are contributing to global climate change.

Read page 26. Use the information from this page to complete the following sentences by underlining the correct answers.

Scientific evidence from Hawaii proves that carbon dioxide levels are **rising steadily** / **staying the same** / **falling**. Drilling into **ice** / **soil** in Greenland and Antarctica suggests that the **weather** / **climate** has gone through cold (glacial) and warm (interglacial) periods. The evidence also suggests that levels of **oxygen** / **carbon dioxide** have gone up and down as part of a natural cycle.

2 Study Figure 11 on page 28 and Figure A below. They show the climate in Stykkisholmur on the west coast of Iceland.

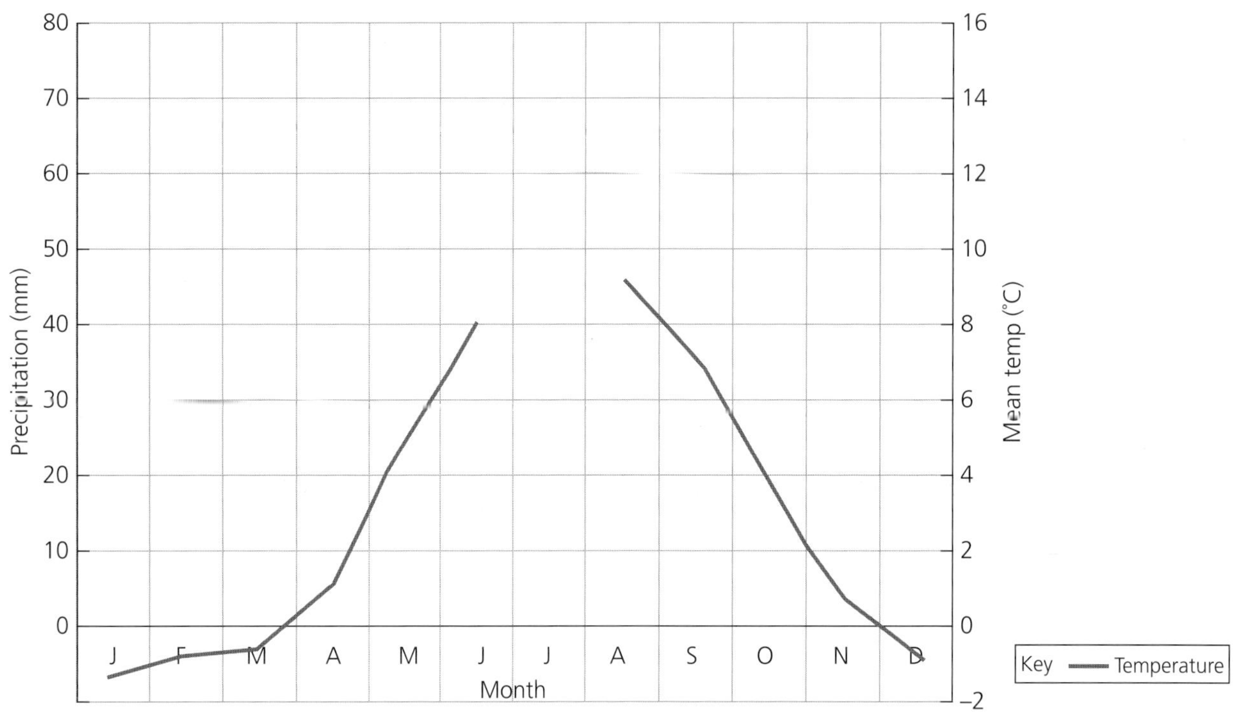

Figure A Climate graph for Stykkisholmur

a) Use the rainfall data below to add bars to Figure A. Complete Figure A by adding the temperature in July and joining the temperature line.

Month	J	F	M	A	M	J	J	A	S	O	N	D
Precipitation (mm)	68	70	72	52	34	40	42	50	58	78	68	70

In July the temperature is 10 °C.

3 Match the following key words to their definitions in the table below.

a) weather **b)** climate **c)** mean temperature

Key word	Definition
	Average temperature taken over a period of time
	The day-to-day condition of the atmosphere
	Weather readings taken over a long period of time

4 Complete the following sentences using answers from the box below.

9	**10**	**11**	**every**	**October**	**December**	**each summer**	**July**

Precipitation (rainfall) in Stykkisholmur is above 30mm in ....

month. The month that receives the highest rainfall is ....

Temperatures are below freezing in the months of January, February, March and

.... The temperature range is .... °C. The highest

temperature is in the month of .... when the temperature is

.... °C.

5 Study Figure 12 on page 28. It shows the annual temperature in Stykkisholmur from 1798 to 2007.

Underline the correct answers in the sentences below.

Annual temperatures in Stykkisholmur have generally **increased** / **decreased** between the years 1800 and 2000. The average pattern (the blue line) shows that between the years 1900 and 1941, temperatures **rose** / **fell** by **–1 °C** / **0 °C** / **1 °C**. Between the years 1941 and 1979 there was a slight **rise** / **drop** in temperatures. However, average temperatures increased again between 1979 and 2003 by **0.2 °C** / **0.8 °C** / **1.5 °C**. Some years have been unusually warm. The warmest year in the last 200 years was in **1941** / **1979** / **2003**.

Review what you have learned

I can:

- understand how scientists are gathering evidence from ice layers about climate change ☐
- draw and interpret a climate graph ☐
- give two details about how Iceland's climate is changing. ☐

How could climate change affect people in the future?

See Core Book pages 32–33

In these activities you will:

- study two cities that are at risk of sea level rise and flooding
- practise the geographical technique of drawing and analysing a bar graph.

1 Explain why New Orleans is at risk of future flooding. Arrange the following phrases into a sequence to help you. Choose some connectives from the box below to join your phrases together.

so	**therefore**	**which means that**	**however**	**because**

- large parts of New Orleans are below sea level
- New Orleans is at high risk of flooding in the future
- the soft sands are drying out
- the city is built on a delta deposited by the River Mississippi
- land in the delta is sinking
- sea levels have been rising by about 2 mm every year
- sea levels may rise by 4 mm a year in the future.

2 Scientists believe that climate change will lead to stormier weather.

a) Use information in Figure 18 on page 33 to complete the empty boxes in the table below.

← Stronger hurricanes

	Category 5	**Category 4**	**Category 3**
Air pressure (mbs)		920–945	
Wind speeds (km/hr)		211–250	
Risk of coastal flooding			Low risk
Damage from wind	Roofs of large industrial buildings blown off		

b) Underline the correct answers in the paragraph below.

Category 5 hurricanes are the **weakest** / **strongest** type of storm. With today's level of CO_2 category 5 hurricanes are rather rare. For example, when the air pressure is 900 mbs there are about **5** / **20** / **30** occurrences each year.

With extra CO_2 in the atmosphere, category 5 hurricanes will become **more** / **less** common. When air pressure is 900 mbs there may be about **5** / **20** / **30** occurrences each year.

The rising threat of flooding in the Thames estuary

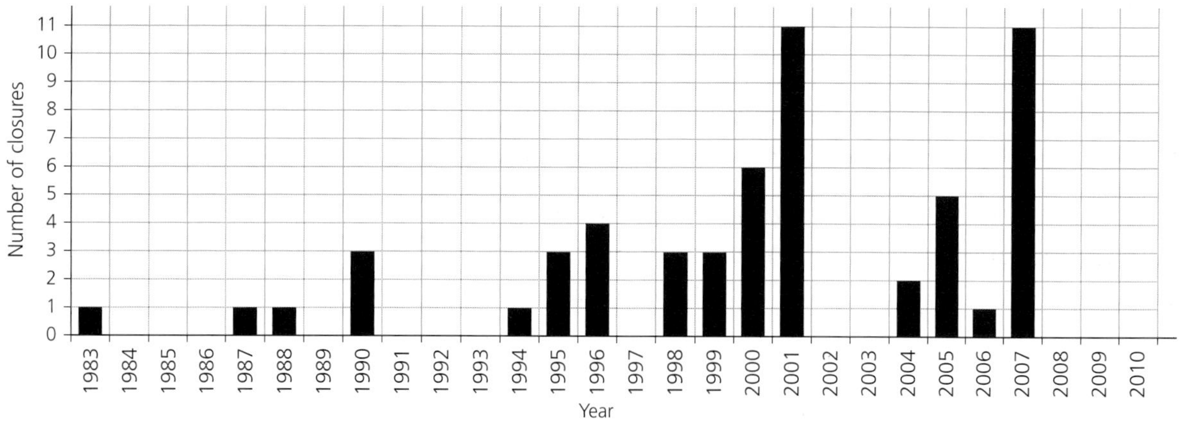

Figure A Number of closures of the Thames Flood barrier to prevent flooding by storm surges

3 Complete the graph of closures of the Thames flood barrier using the data below.

Year	1992	1993	2002	2003	2008	2009	2010
Number of closures	1	5	2	8	3	5	5

4 Complete the sentences below. Fill in the gaps and underline the correct answers.

The Thames barrier was closed .... times in the first ten years (1983–1992). It was then closed **35** / **38** / **41** times in the next ten years (1993–2002). The largest number of closures in any one year was in .... and ....

The number of closures each year has been **going up and down** / **steadily decreasing** / **staying the same**, and there were significantly **more** / **fewer** closures in the 2000s than in the 1980s.

✎ **5** Do some research into one of the following cities. Explain why your chosen city is at risk from climate change.

a) Mumbai (Bombay), India

b) Dhaka, Bangladesh

c) London, UK

Review what you have learned

I can:

- draw and analyse patterns from a graph ☐
- name two places in the world that are at risk of tropical storms and floods ☐
- explain the connection between greenhouse gas emissions, global warming and increased risk from weather hazards. ☐

How can we create a low carbon future?

See Core Book pages 38–41, 44

In these activities you will:

- describe how Europe could create 'green' energy
- consider conflicting attitudes about low carbon futures
- explain why the BedZED housing development is described as a sustainable community.

1 Read about the solar furnace, shown in Figure 30, on pages 38–39. Use this information to add one number to each of the statements below:

a) There are .... mirrors in the solar furnace.

b) There are .... homes in Seville which get electricity from the solar furnace.

c) The estimated cost of building a 'super-grid' to bring electricity generated by solar energy from the Sahara to Europe is ....

2 Study Figure 32 on page 39. Use information from this map to complete the following sentences by underlining the correct answers. You may also need to use an atlas or Google Earth.

Two / **three** / **four** European countries already make 10,000 mega watts of electricity from wind energy. These include **Italy** / **France** / **Germany**. The UK produces **0–2,000** / **2,000–4,000** / **4,000–6,000** mega watts from wind energy. Iceland, Scotland, and **France** / **Germany** / **Norway** all have coastlines with high potential for wave power.

3 a) Read Figure 33 on pages 40–41. Concentrate on 'energy conservation and efficiency' and growing 'biofuels'.

b) Use this information to complete the following table.

Method of creating a low carbon future	Energy conservation and efficiency	Growing biofuels
Who supports this method?		A spokesperson for the biofuel industry
What are their arguments?	There are two simple ways families could reduce energy. These are: 1. 2.	
Who objects to this?	Young families on low incomes	
Why do they object?		Biofuels are causing poverty and hunger because …

Case study: BedZED – a sustainable community

Key word: Something is **sustainable** when it meets people's needs now and in the future.

Something is **socially sustainable** when it provides benefits for families, health, education and leisure.	Something is **environmentally sustainable** if it conserves air quality, habitats and wildlife.	Something is **economically sustainable** when it creates jobs, income and spending.

4 Below (A–I) are features of the BedZED sustainable community.

a) Colour socially sustainable features in yellow.

b) Colour economically sustainable features in blue.

c) Colour environmentally sustainable features in green.

Some statements may be sustainable in more than one way. You could choose to colour the statement in more than one colour.

A BedZED uses solar energy and woodchips to heat the homes and provide power.	**D** There is a crèche, youth group and community centre. These provide facilities for people of all ages.	**G** Roof gardens soak up rainwater so less flows into storm drains.
B BedZED is a carbon-neutral development – the homes do not add any extra carbon dioxide emissions into the atmosphere.	**E** There is a car pool for residents to reduce the number of car journeys.	**H** The buildings are very well insulated which conserves heat energy in the homes.
C There is a mix of tenures including fifteen homes at affordable rent for social housing.	**F** Where possible, the buildings have been built from natural, recycled or reclaimed materials.	**I** There are large windows which face south to collect heat energy from the Sun.

✎ **5** Choose **three** statements (from B–I). Explain why each is sustainable. One has been done for you.

A BedZED uses solar energy and woodchips to heat the homes and provide power. This is environmentally sustainable because less fossil fuel will be burned so BedZED makes a smaller contribution to climate change.

Review what you have learned

I can:

- explain the point of view of two different groups of people ☐
- describe three features that make BedZED sustainable. ☐

Living in an Active Zone

What happens at constructive plate margins?

See Core Book pages 46–47, 51

In these activities you will:

- describe the features of a constructive plate margin.

1 Study page 46 and the key words below. Tick (✓) the correct definition for each term.

a) Destructive plate margins are

where two plates pull away from each other ☐

where two plates meet and one is destroyed as it is pulled under the other ☐

b) Constructive plate margins are

where two plates are pulling apart and new crust is formed ☐

where two plates meet and do not collide ☐

2 Figure 3 on page 46 shows landforms of a constructive plate margin in Iceland and beneath the North Atlantic Ocean. Figure A below shows the same landforms.

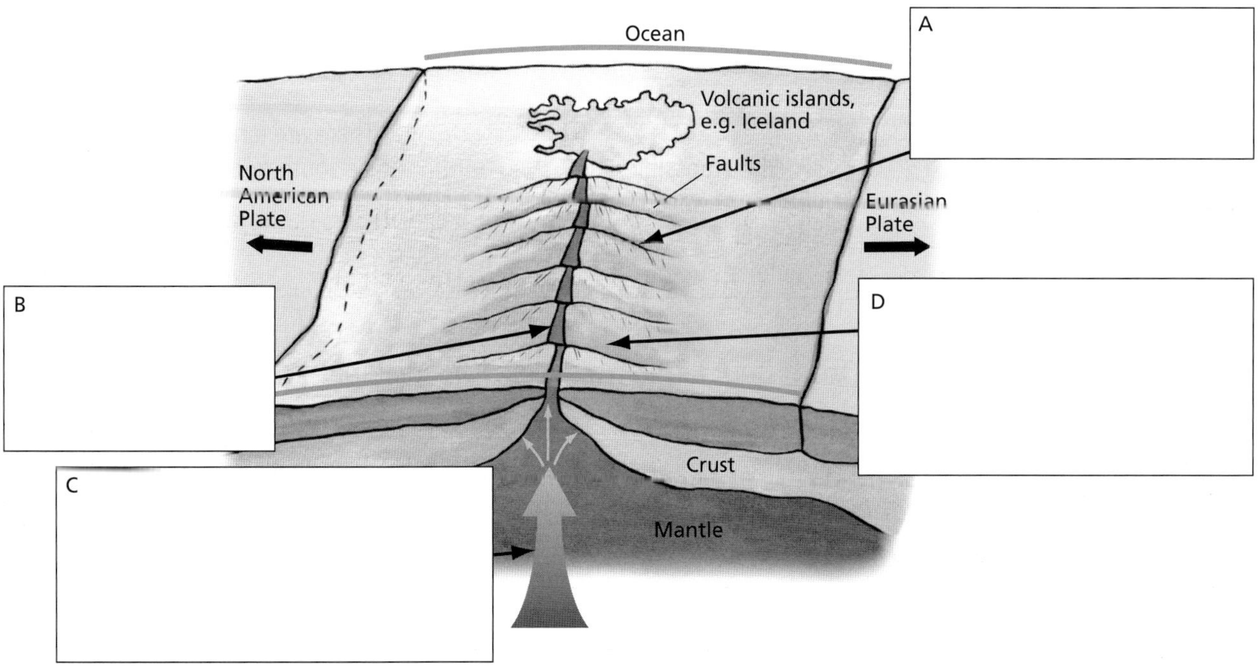

Figure A Landforms of a constructive plate margin

a) Add labels to boxes A and B that describe the features.

b) Add annotations to boxes C and D to describe processes occurring here.

3 Read page 47 and study Figure 4. Underline the correct answers in the sentences below.

Iceland is a country being torn into **two** / **three** / **four** pieces. A giant fracture has **opened up** / **closed** the Earth's surface. As this has widened the land at the fracture has **lifted up** / **sunk**, forming a **rift valley** / **oceanic ridge** with steep cliffs (escarpments) running along both sides.

4 Iceland has a number of very large **shield volcanoes**. Study Figure 6 on page 47. Use words from the box below to help you write labels in the boxes around Figure B describing the features of a shield volcano.

steep height	**gentle slopes**	**narrow area**
steep slopes	**tall in height**	**lava spreads easily**
lava runs down in channels	**wide area**	**low height**

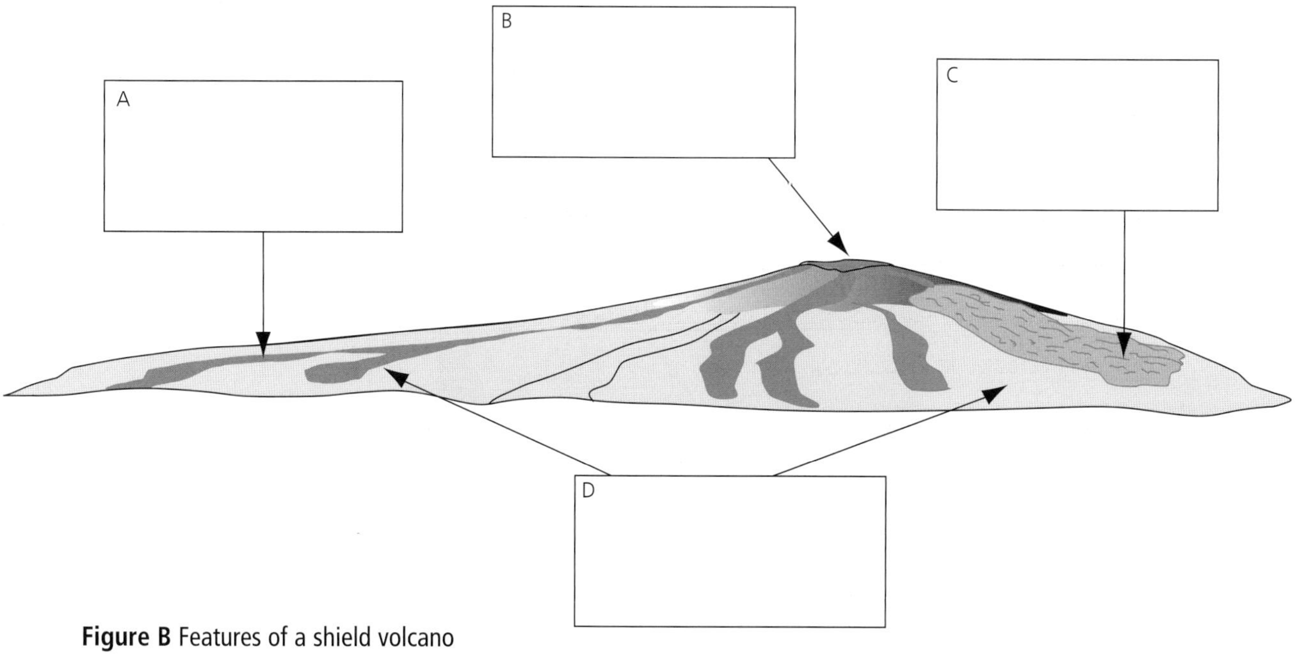

Figure B Features of a shield volcano

5 Read page 51. It describes some of the effects volcanic eruptions can have in Iceland.

Decide whether each statement describes the effect of the volcano's eruption on people, or the environment, or both.

Case study	People (P) / Environment (E) / Both (B)
Huge clouds of ash cause disruption to air traffic across Europe.	
The weight of ash deposits causes the roofs of buildings to collapse.	
Lava flows destroy buildings, roads and trees.	
Ash contaminates water supplies and smothers crops.	
Livestock can be killed due to eating grass contaminated with ash.	
Crops fail to grow and this results in famine.	

Review what you have learned

I can:

- describe the processes that take place at constructive plate margins ☐
- label three features of a shield volcano. ☐

What happens at destructive plate margins?

See Core Book pages 48–49

In these activities you will:

- explain the processes taking place at destructive plate margins
- describe the features of a strato-volcano.

1 Study Figure 7 on page 48 and Figure A below. They show the landforms produced by subduction of an oceanic plate beneath a continental plate.

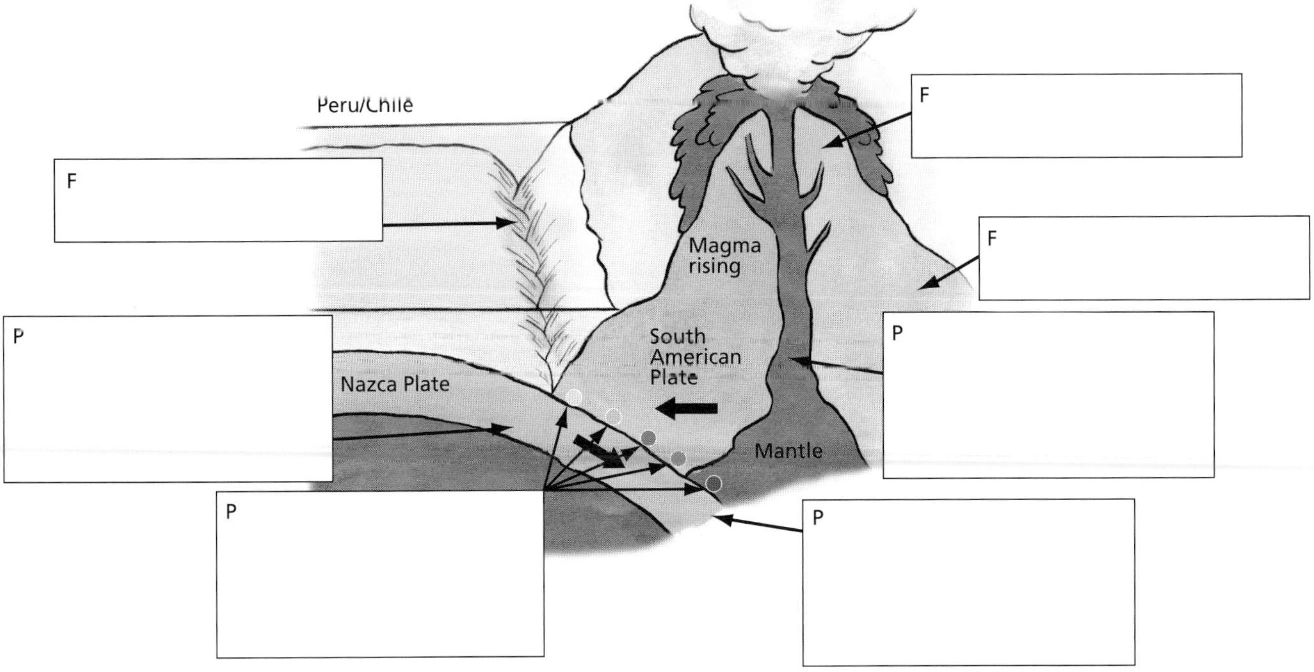

Figure A Landforms of a destructive plate margin

a) Add labels to the boxes that describe the features. These boxes are marked with an F.

b) Add labels to the boxes that describe the process taking place. These boxes are marked with a P.

2 Study page 48 and Figure 7. Number the following chain of events to explain what happens at the subduction of the Nazca Plate in South America.

Chain of events	Number order (1–5)
This causes intense friction and pressure increases.	
Magma rises forming volcanoes and a chain of fold mountains in the Andes.	
The Nazca Plate is being pulled down into the mantle.	
Due to the intense friction, the temperature in the mantle increases.	
The pressure becomes too great and there is a sudden jolt, causing an earthquake.	

3 Study page 49 and Figure 9 to describe what happens when two oceanic plates move towards each other. Complete the description by using words from the box below.

Caribbean	**crust**	**trench**	**sediment**	**melts**
subducted	**magma**	**North American**	**island arc**	

In the Caribbean the .... plate is being .... under the .... plate. This has caused an oceanic .... to form. The subduction of the oceanic crust drags down .... and seawater from the sea bed. This mixture .... due to the intense heat, causing .... to rise upwards into the Caribbean .... This forms a chain of mountains called an ....

4 Study Figure 10 on page 49. Use appropriate words from the box below to help you label the boxes around Figure B to describe the main features of a **strato-volcano**.

gentle slopes	**steep slopes**	**explosive eruptions**
layers of ash and lava	**high summit**	**lava spreads easily**

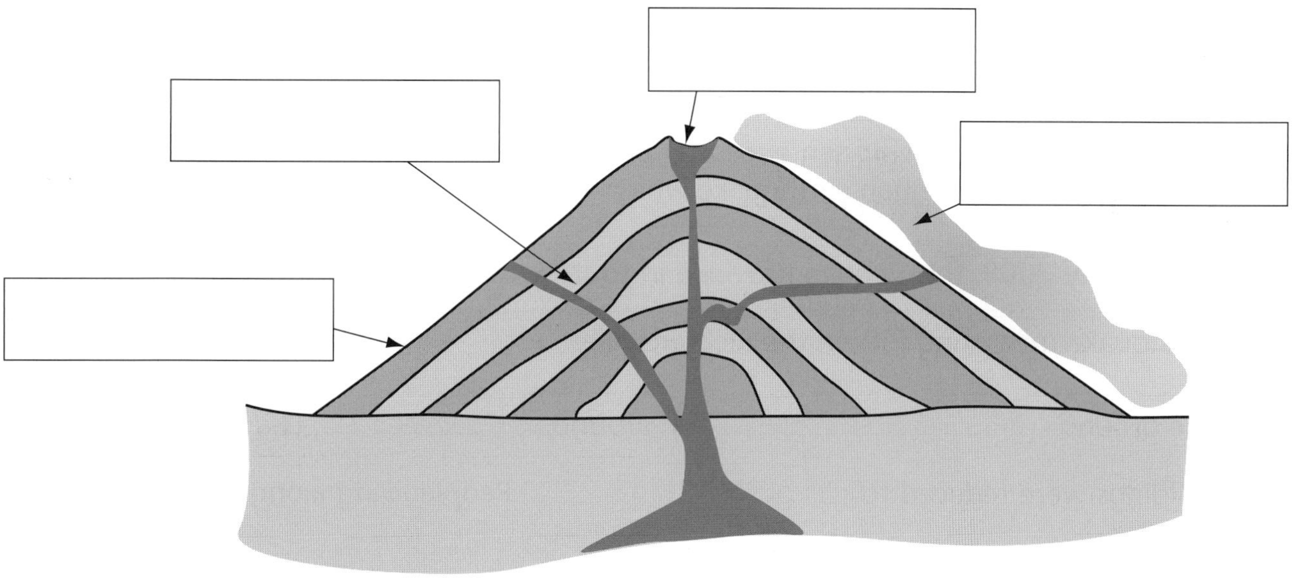

Figure B Features of a strato-volcano

Review what you have learned

I can:

- explain the chain of events that take place at destructive plate margins ☐
- label three features of a strato-volcano. ☐

Volcanic and earthquake hazards

See Core Book pages 50, 54–55

In these activities you will:

- identify the primary hazards associated with an earthquake or volcanic eruption
- describe the change in the environment following an earthquake or eruption.

1 Match the key words to the definitions in the table below. Read the text on pages 50 and 54 and refer to the glossary pages 133–136.

a) primary hazards **c)** pyroclastic flow **e)** lahars **g)** liquefaction

b) secondary hazards **d)** tephra **f)** aftershocks

Key word	Definition
	Fragments that are thrown from a volcano during an eruption
	A process that turns soil to jelly which can take place during a strong earthquake
	A flood of water and volcanic ash or a mudslide down the slope of a volcano
	Events that can cause risk to life following an earthquake or eruption
	An earthquake that occurs minutes or days after a major earthquake
	Events that cause risk as a direct result of an earthquake or volcano
	A landslide of hot ash, rocks and gases from a violent eruption

2 Read the information on pages 54 and 55.

a) Complete the following sentences in your own words to describe the **primary hazards** of the Sichuan earthquake which struck China in 2008.

On 12 May 2008, the Sichuan province in south-west China was struck by a strong earthquake measuring 7.9 on the Richter scale.

The ground .... and around .... lost their lives.

Buildings were reduced to .... People living along the river valleys found that ....

In Beichuan ....

....

b) Describe how the town of Beichuan was affected by **secondary hazards** in the days and week after the earthquake. How did the Sichuan earthquake affect people and the environment? Identify the different effects of the earthquake by placing a tick (✓) in one or more columns in the table below.

Effects of the earthquake	Social	Economic	Environmental
Communication links were destroyed.			
375,000 people were injured.			
Rice crops were destroyed or harvested early.			
Trains were derailed, landslides blocked roads.			
4.8 million people were made homeless.			
Power cables were brought down.			
Powerful aftershocks continued for months.			
Small fires burned in Sichuan.			
Increased threat of diseases such as cholera.			
Industries were shut down as many workers were killed.			
The town of Beichuan was rebuilt elsewhere.			

3 Study Figure 20 on page 55. Complete the sentences below by filling the gaps.

Another example of a **secondary hazard** is when an earthquake struck .... in the Indian Ocean on 26 December 2004. This earthquake, which measured .... on the Richter scale of magnitude, caused a tsunami which killed .... people. This proves that the follow-on effects of an earthquake can sometimes be far worse than the **primary hazards.**

Review what you have learned

I can:

- describe two primary and two secondary effects of the earthquake in Sichuan ☐
- classify the impact upon people, the environment and the economy. ☐

Reducing risk

See Core Book pages 58–59, 62

In these activities you will:

- describe how volcanoes are monitored
- describe the measures used to reduce the risk of earthquakes.

1 Scientists are consistently monitoring and trying to predict when volcanoes will erupt. Study pages 58 and 59. Match up the signs that a volcano may erupt with what it tells us about a volcano by drawing a line between them.

Signs that a volcano may erupt	What does this tell us?
There are lots of small earthquakes …	magma is moving causing the dome of the volcano to become steeper and dislodge loose rock.
The ground may deform, for example a bulge may grow on the volcano …	gases are escaping as the magma gets close to the surface.
Rockfalls occur on the slopes of the volcano …	the magma chamber is inflating like a balloon.
Sulphur dioxide and CO_2 are released from vents in the volcano …	magma is moving up and this pressure causes rocks to crack and split.

2 Study the data in Figure 27 on page 59 which gives information leading to a pyroclastic flow.

a) Complete Figure A below by adding scattergraph plots to show the number of earthquakes and rockfalls on each date.

b) Describe what is shown by your completed scattergraph.

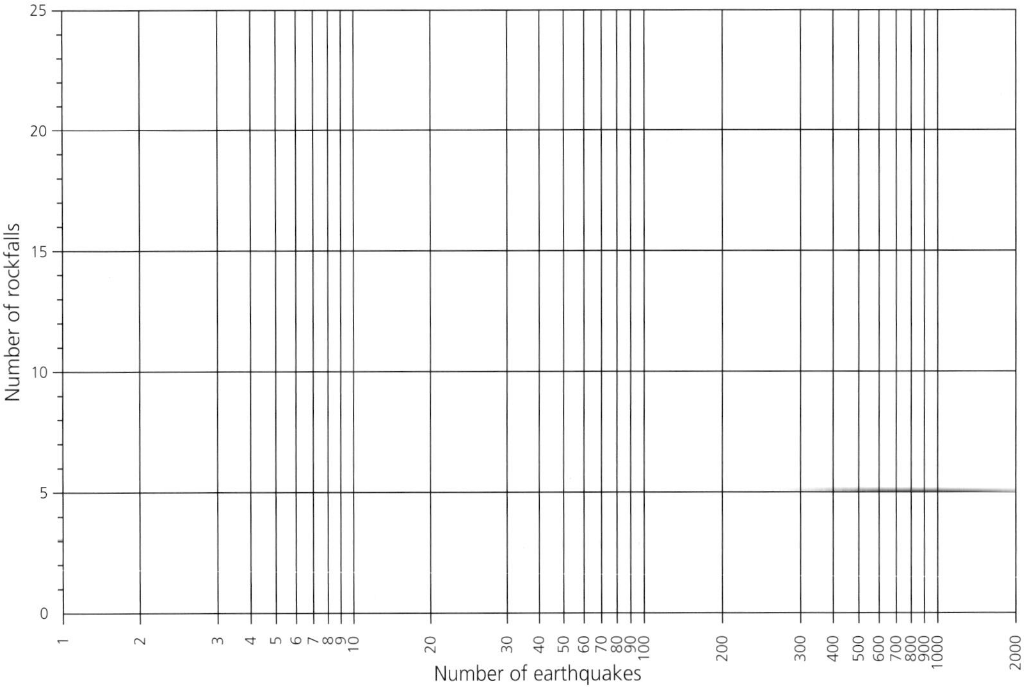

Figure A

3 LEDCs are less prepared than MEDCs for volcano and earthquake disasters. Put the numbers 1–8 below in the correct places on the Venn diagram in Figure B.

1 Lots of money invested to reduce risk	**2** Technology used to monitor changes
3 Reliance on international communities	**4** Less prepared due to little money
5 Warning systems to people living in vulnerable areas	**6** An emergency planning system in place
7 Buildings, roads and bridges designed to withstand earthquakes	**8** Earthquake education and drills

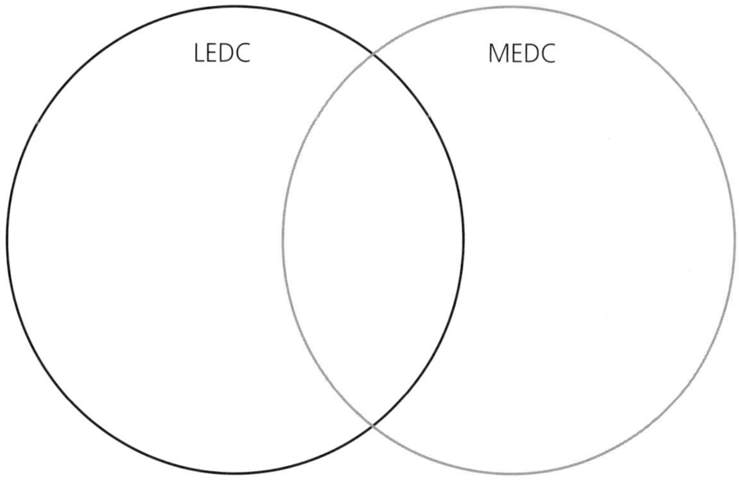

Figure B

4 Study Figure 30 on page 62. It shows an earthquake-proof building in an **MEDC**. Explain how each feature in the table below will reduce the impact of an earthquake. One has been done for you.

Feature of building	**How this will reduce the impact of an earthquake**
Shock absorbers in the foundations	This will absorb the energy in the seismic waves so there is less of a shake impact to the building.
Cross bracing	
Roof covering the area outside	
Several emergency staircases and exits	
Weights on the roof	
Large open areas	
Deep foundations in solid rock	

Review what you have learned

I can:

- describe three ways that volcanoes are monitored ☐
- describe what measures are taken to reduce the effects of earthquakes. ☐

Changing Populations

Investigating patterns of migration in South Africa

See Core Book pages 70–73

In these activities you will:

- explain why people are moving from the countryside (**rural** areas) to cities (**urban** areas)
- consider the impacts of migration on both urban and rural areas.

1 Match the following key words to their definitions in the table below.

a) Economic migrant
b) Natural population change
c) Net migration
d) Poverty line
e) Brain drain
f) Remittances
g) Circular migration

Key word	Definition
	The difference between the birth and death rates
	The movement and return of people between rural and urban areas
	The difference between the number of people moving into an area and the number of people moving out of an area
	A person who moves to live somewhere else to earn money
	When people live below a level of income and are said to be poor
	Money sent by migrant workers to support their families who have remained at home
	When an area loses highly qualified workers due to emigration

2 Limpopo is South Africa's most rural province. Many people move from here to Guateng, which is the neighbouring province. We need to investigate why this is.

a) Study Figure 7 on page 70. Use the information from this map and other information on pages 70–73 to complete the fact boxes in Figures A, B and C.

The province of Gauteng has an urban population of .... per cent.	The province of Limpopo has an urban population of .... per cent.
Gauteng has three cities of more than 1 million including	The largest city in Limpopo has a population of ....
The number of people in Gauteng with no schooling is	The number of people in Limpopo with no schooling is ....

Figure A Comparing life in Gauteng to Limpopo

Pull factors	Push factors
Housing in Gauteng is	Housing in Limpopo is ....
Income in Gauteng is	Income in Limpopo is ....
The percentage of people living below the poverty line in Gauteng is lower than that of Limpopo at	The percentage of people living below the poverty line in Limpopo is ....

Figure B Why are people moving from Limpopo to Gauteng?

Negative impacts of the rural to urban migration on Gauteng include	Negative impacts of the rural to urban migration on Limpopo include ....
Positive impacts of the rural to urban migration on Gauteng include	Positive impacts of the rural to urban migration on Limpopo include ....

Figure C What are the impacts of migration? (Read pages 72–73)

Review what you have learned

I can:

- describe three push and three pull factors ☐
- use South Africa as a case study of migration, giving three facts. ☐

How are rural populations changing?

See Core Book pages 74–75

In these activities you will:

- examine the reasons for people moving to and from the UK countryside.

1 Read pages 74 and 75. Match the following key words to their definitions in the table below.

a) counter urbanisation **b)** commuting **c)** tele-working

Key word	Definition
	The daily or weekly movement of people to and from work
	Jobs where workers work from home often using the internet to communicate
	The process by which people move from cities to live in the countryside

2 Study Figure 14 on page 74.

a) A lot of people now **telework** from homes in the countryside. Give **three** changes that have made it possible for more people to live and work in the countryside. One has been started for you.

i) More powerful computers so ....

....

ii) ....

....

iii) ....

....

b) Give **two** changes that make it possible to easily visit the countryside for leisure activities.

i) ....

ii) ....

3 The process of counter urbanisation means there are now social differences between rural and urban regions in the UK. In this activity you will examine differences in population structure and occupation.

a) Study Figure 16 on page 75. Complete the following table. Some answers have been completed for you.

Population	Shrewsbury (urban)	connective	Clun (rural)
Aged 60–74	14%	whereas	22% are 60–74 which is 8% higher.
Aged 20–29			
Aged 10–19	14%		

b) Complete the following conclusion about population structure.

There are more .... people in rural Shropshire than in urban areas. There are a lot fewer people aged .... in rural areas, perhaps because they have left to look for work.

c) Use Figure 17 on page 75 to complete the following conclusion about occupation:

There are a lot more .... people in rural Shropshire than in urban areas. However, there are slightly fewer highly skilled occupations such as ....

4 Some people choose to move to rural areas like Shropshire while others prefer to move out of the countryside and into cities. Use information from pages 74–75 to add push and pull factors to the table below.

	Push factors	**Pull factors**
Families or retired professional people moving out of larger towns into the countryside		
Young adults moving out of the countryside into larger towns		

Review what you have learned

I can:

- give three reasons why people are able to telework ☐
- describe three differences between urban and rural populations ☐

Why is it important for girls to complete their education?

See Core Book pages 81, 123

In these activities you will:

- show you know why it is beneficial for girls to complete their education
- practise the geographical technique of drawing a shaded (choropleth) map.

In some parts of South Asia, girls miss a lot of school. This means a lot of young women cannot read and write properly. This causes serious problems. For example, the child of an uneducated mother is twice as likely to die before the age of one as a child whose mother has a full education.

1 Study Figure 24 on page 123 carefully. It shows the advantages that an educated mother has compared to a mother who was not able to finish school.

a) Close the textbook! Now complete Figure A below by adding words from the box in each blank space. You may use words more than once.

better	**lower**	**earlier**	**higher**	**fewer**	**later**	**sooner**	**less**	**worse**

b) When you have finished, check your answer by looking at Figure 24 on page 123 again.

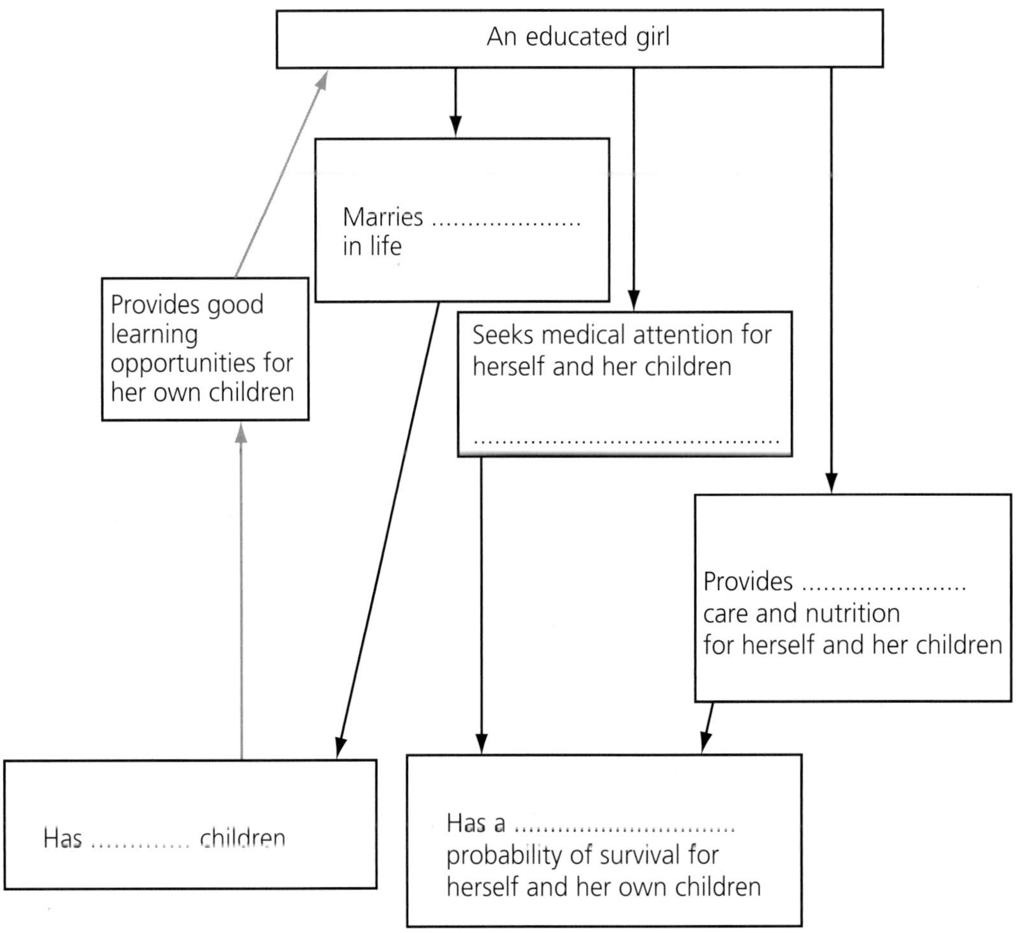

Figure A The future advantages of educating girls

2 Use Figure A to complete each of the following statements.

- An educated mother spots the early signs of ill health in her child so …
- An educated mother understands the importance of a balanced diet so …
- An educated mother recognises the importance of a full education for her daughter so …

3 Use the data in Figure 25 on page 81 and Figure B below to make a map showing female literacy in India.

a) Choose four colours for the key. Use the lightest colour for the smallest numbers.

b) Colour each state on Figure B using your key. For example, Andhra Pradesh has female literacy of 50.4% so it fits into the range 45%–59.9%.

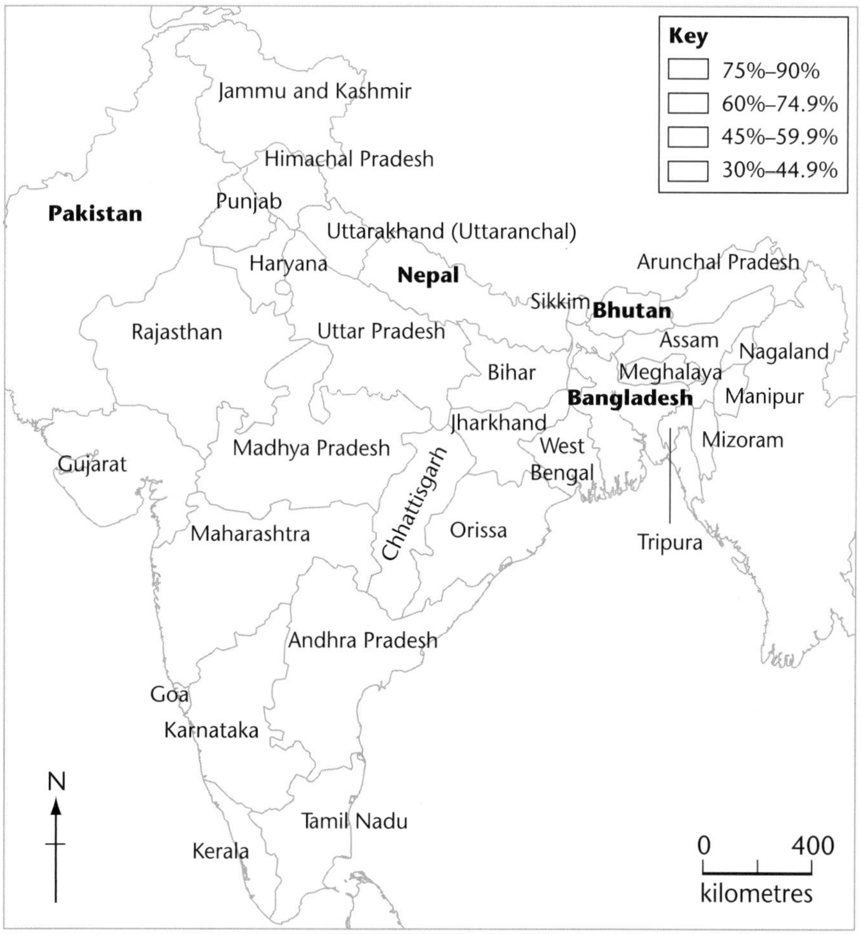

Figure B Outline map of the states of India

4 Compare your finished map to Figure 24 on page 81.

a) What are the similarities between the two maps?

b) What does this tell you about the importance of education for girls?

Review what you have learned

I can:

- give three reasons why girls should complete their education ☐
- draw a shaded (choropleth) map and analyse the pattern that it shows. ☐

Investigating patterns of fertility in India

See Core Book page 81

In these activities you will:

- practise the geographical technique of drawing a scattergraph
- make predictions and investigate relationships.

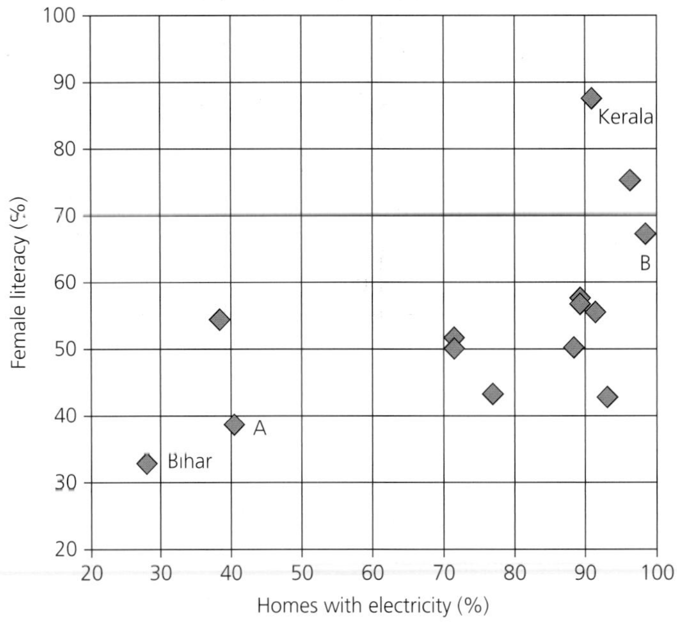

Figure A Investigating the relationship between homes with electricity and female literacy in Indian states

Figure A is a scattergraph. A clear pattern on a scattergraph may indicate that two sets of data are linked in some way.

1 Figure A has been drawn using data from Figure 25 on page 81. The graph plots data for the first fourteen states in the table from Andhra Pradesh down to Madhya Pradesh.

a) Use information in the table to identify the states A and B.

State A is ....

State B is ....

b) Complete the following sentences about Figure A.

States which have few homes with electricity have **low** / **medium** / **high** female literacy. An example is .... States which have many homes with electricity have **low** / **medium** / **high** female literacy. An example is ....

c) Suggest a reason for the pattern shown in Figure A. Tick (✓) one of the following answers to show the most likely reason:

- ☐ Homes with electricity have better lighting so children can do their homework.
- ☐ The poorest homes have no electricity. Families in these homes are so poor they cannot afford to send their daughters to school.
- ☐ Some Indian states do not have enough money for either schools or electricity connections.

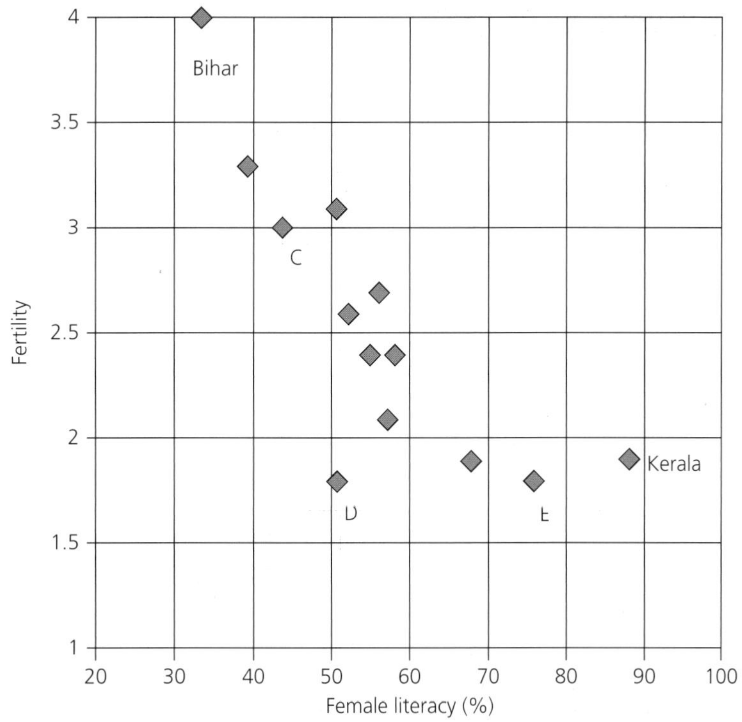

Figure B Investigating the relationship between female literacy and fertility

2 Figure B has been drawn using data from Figure 25 on page 81. The graph plots data for the first fourteen states in the table from Andhra Pradesh down to Madhya Pradesh.

a) Use the table to identify the states C, D and E.

State C is ....

State D is ....

State E is ....

b) Complete the graph by adding ten more states from Figure 25 on page 81.

c) Complete the following sentences about Figure B.

States which have low female literacy have **low** / **medium** / **high** fertility. An example is ....

States which have high female literacy have **low** / **medium** / **high** fertility. An example is ....

Review what you have learned

I can:

- draw a scattergraph ☐
- make predictions and use a scattergraph to investigate relationships. ☐

Globalisation

Why do MNCs have factories in different parts of the world?

See Core Book pages 92–93

In these activities you will:

- investigate the location of a multi-national company
- consider why MNCs have factories in different parts of the world.

1 Study Figure 5 on page 92. You may also need to look at an atlas or on Google Earth.

a) Underline the correct answers in the following sentences.

Nokia has **three** / **four** / **five** research and development (R&D) laboratories in the USA. It has another **five**/ **six** / **seven** R&D laboratories in Europe and one in **Australia** / **Japan**. These are all developed countries **north** / **south** of the Brandt line. There is one other R&D laboratory in **Africa** / **India** / **China**. In total there are **nine** / **ten** / **eleven** R&D laboratories north of the Brandt line and **one** / **two** / **three** south of the Brandt line.

b) Now look at the distribution of production plants (factories) on Figure 5. Use your answer to activity 1 a) to help you structure a description of the distribution of these factories.

c) Which developing region of the world has:

i) sales offices in most countries? ....

ii) hardly any sales offices? ....

d) Suggest a reason for the pattern of sales offices.

e) Complete the following conclusion by adding your own word to the gaps in the sentences below.

Nokia's head office is in .... Most of Nokia's R&D laboratories are in .... countries whereas over half of its factories are in .... countries.

2 Study Figure 8 on page 93. Use the data in this table to complete Figure A opposite by adding lines for:

- Germany
- China
- India

Use a different colour for each country and don't forget to add your lines to the key.

Figure A Nokia's sales (2003–2007)

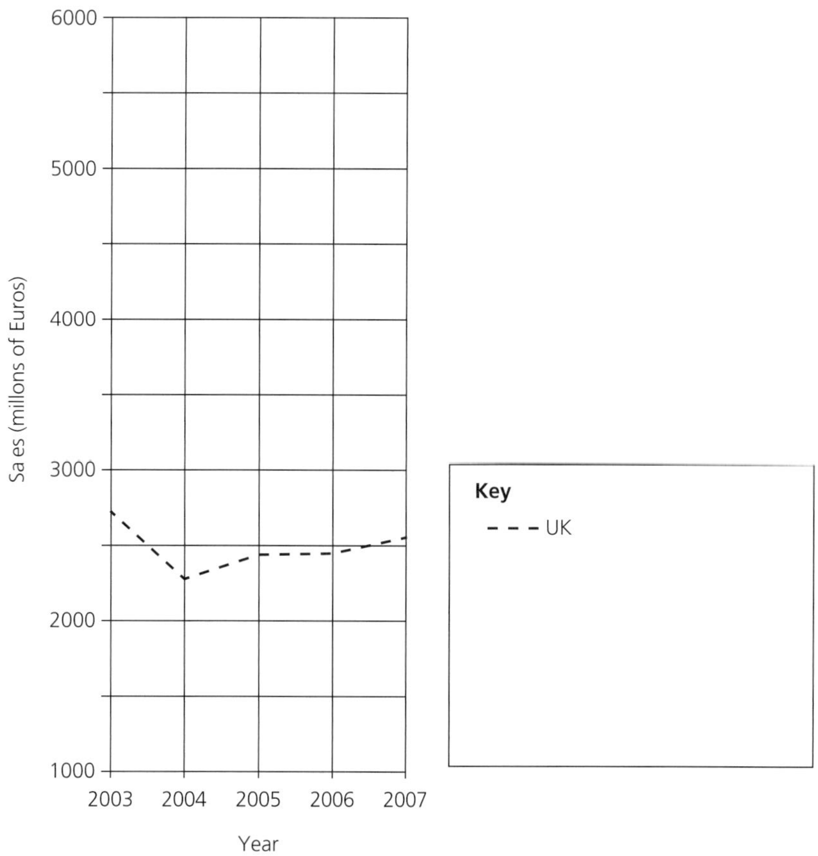

3 So why do MNCs have factories and offices in different parts of the world?

a) Work in pairs. Discuss the information on pages 92–93.

b) Use your ideas to complete the matching exercise below.

During 2003–2007 sales of Nokia's products were growing rapidly in India so …	… it made sense to keep wage bills down by making more products in India and China.
Wages in factories in India in 2007 were much lower than in Germany or the UK so …	… it made sense to keep most of its R&D laboratories in countries that had a large highly skilled workforce.
In 2007 building costs in south Asia were much lower than in Europe so …	… it made sense to make more phones in south Asia to keep down the cost of transporting the finished product to the shops.
Nokia is a high tech company that needs to constantly create new products so …	… it made sense to keep costs down by renting properties in India and China rather than in Finland or Germany.

Review what you have learned

I can:

- describe the global distribution of Nokia's plants ☐
- give four different reasons why MNCs have factories in different parts of the world. ☐

Why do people move within the EU?

See Core Book pages 98–99

In these activities you will:

- practise drawing a block graph and analysing what it shows
- explain why people might move from one EU country to another.

1 Use Figure 15 on page 98 to complete the following sentences about the EU by underlining the correct answers.

The EU is an example of a trade **bloc** / **group**. In 2003 there were **fourteen** / **fifteen** / **sixteen** countries in the EU including the UK, France and **Russia** / **Germany**. In 2004 the number of countries in the EU was **enlarged** / **reduced** by eight countries such as Hungary and **Portugal** / **Poland** which were in **central and eastern** / **southern** / **northern** parts of Europe. The last two countries to join were Bulgaria and **Romania** / **Ukraine**. There are now **25** / **26** / **27** members of the EU.

2 Study Figure 16 on page 98 and Figure A below.

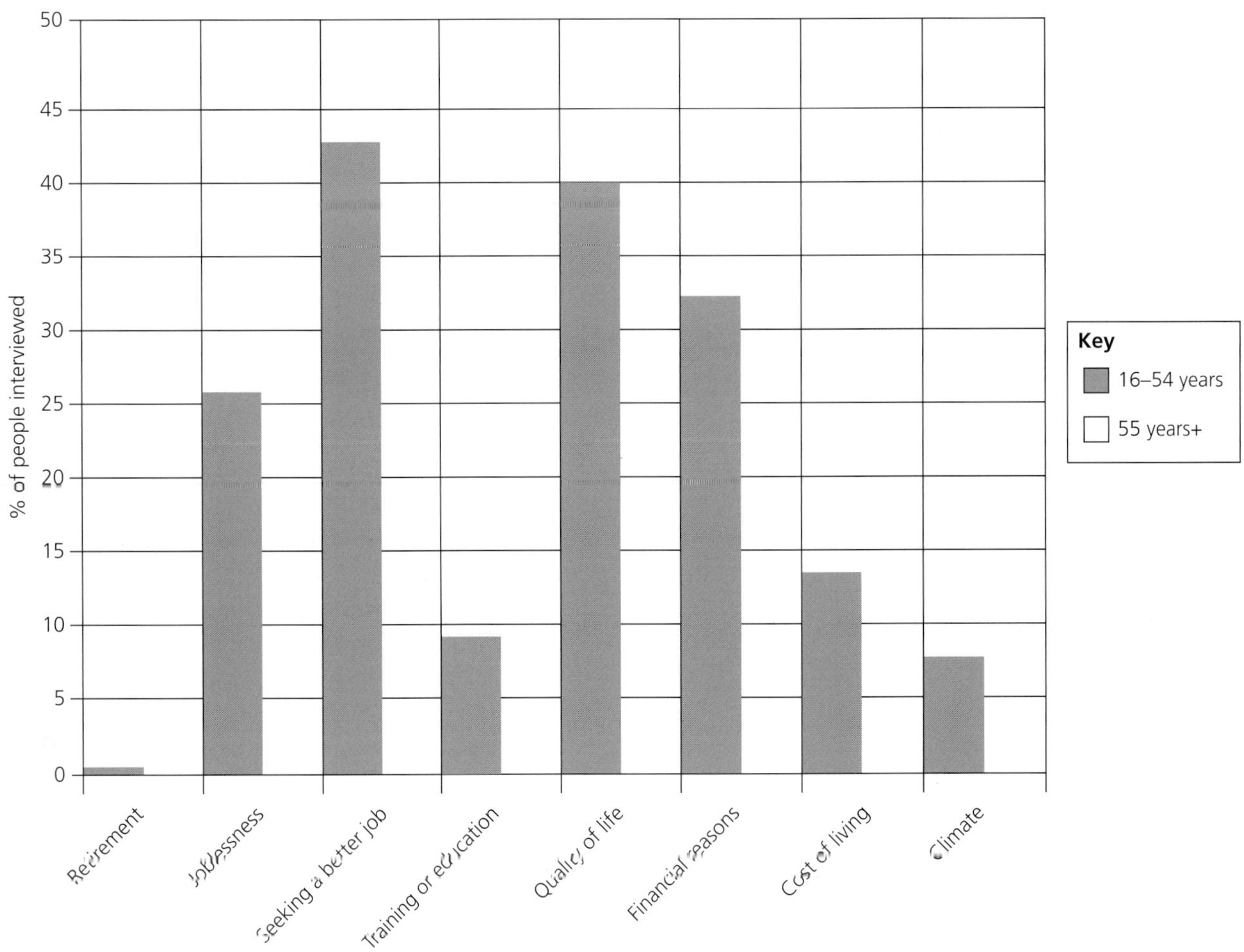

Figure A Reasons given by migrants who moved to Spain from other EU countries

a) Use data from Figure 16 to add bars for people in the age group 55 years+ to complete Figure A.

b) Complete the key for Figure A.

c) Think about the reasons given by these two age groups:

i) Give one similarity: ....

ii) Apart from retirement, give one difference: ....

✎ **d)** Suggest why someone from the UK might move to Spain when they retire. Use the ideas below to support your answer and extend your answer using connectives.

Starting point	Connectives	Ideas to help you elaborate your answer
1 Spain has a warm climate …	so … therefore … whereas …	What is the climate like in Spain compared to the UK in the winter months? How might this affect heating bills? How might a better climate affect your health?
2 The cost of living is …		People who have a pension have limited incomes. What might it cost to shop for food, eat out or heat your home in Spain compared to the UK?

3 Study Figure 17 on page 98.

a) Use information from this article to complete the matching exercise below.

Some migrants do work that is dirty, difficult and dangerous such as …	… fruit picking, road repairs or jobs in the tourist industry such as restaurant work.
Some migrants do service industry jobs (that are usually low paid) such as …	… jobs in construction.
Some migrants do jobs that are seasonal (where workers are needed only for a few months) such as …	… doctors, surgeons or jobs in the information technology sector.
Some migrants do jobs in highly skilled or knowledge intensive industries such as …	… cleaning, childcare or working in fast food outlets or pizza delivery.

✎ **b)** Use information from Figure 17 to explain why many people think that migrant workers benefit the country to which they move. Use ideas from 3a) to extend the answers below:

Many local people do not want to do work such as …

These jobs are unpopular because …

However, it is important that these jobs are done because … For example …

Review what you have learned

I can:

- draw a block graph and analyse its main features ☐
- explain why people move within the EU. ☐

Globalisation

See Core Book pages 89, 100–102

In these activities you will:

- describe the process of globalisation
- consider the benefits and problems of globalisation for India.

1 Read each statement in the table below. Decide whether it is a potential cause (C) or effect (E) of the process of globalisation. One has been done for you.

	C/E		C/E
The invention of the internet		Faster, larger airplanes	C
We can buy tropical fruits in the UK		Satellites that improve communications	
Growing demand for expensive goods in NICs such as India and China		People migrate within Europe to find work	
Electronic transfer of money		People watch Bollywood films in the UK	
The growth of mobile phone systems		Companies open factories or offices in different countries	

2 Study Figure 19 on page 100.

a) Complete the following description of the graph by underlining the correct answers.

The number of mobile phone customers in Japan was **10** / **20** / **50** million in 1996. Numbers **rose steadily** / **fluctuated** / **fell** reaching just under **50** / **100** / **150** million in 2006. In 1996 the number of mobile phone customers in India was **lower** / **higher** than in Japan. Since 2001 numbers have increased **steadily** / **rapidly**. The number of mobile phones in India overtook the number in Japan during **2004** / **2005** / **2006**.

b) How does India benefit from the growth in mobile phones? Complete the heads and tails exercise below.

Incomes in India are rising …	… so creative industries such as gaming, film and advertising grow.
India produces thousands of IT and software graduates each year …	… so businesses can make deals, benefit from new ideas and training.
Mobile phones allow easy sharing of ideas and culture …	… so people can afford more consumer goods like mobile phones.
Mobile phones encourage networking …	… so there are skilled people to work in high tech industries.

3 Do you think that Tata is good for India?

a) Read the statements below about their new car, the Nano. They are based on the information on page 102.

b) Decide where each statement fits best to complete the table below. One has been done for you.

1. Poor families ride on scooters which can be dangerous. The Nano means that families will have a safe form of transport.

2. The Nano is cheap. It will be affordable to millions of Indian families.

3. The Nano produces very little pollution compared to other vehicles in Indian cities.

4. The Nano uses very little fuel.

5. There is too much traffic in Indian cities. Tata's new car will make traffic jams worse.

6. The Nano is clean, but there will be so many of them that air pollution will get worse.

7. People in West Bengal did not want Tata's new factory. They would have lost their land but would not have got jobs in the factory.

8. The Nano will be sold to other developing countries which will be good for India's pattern of trade.

9. As Indians get wealthier the demand for new cars will grow. So, even more jobs will be created.

	Benefits for India created by Tata's success	**Problems for India created by Tata's success**
Economic		
Environmental		
Social	1	

4 Use Figure 20 on page 101 to describe how workers in Europe will benefit if the Tata Nano is successful. Include facts and place names.

Review what you have learned

I can:

- give three causes of globalisation ☐
- give three effects of globalisation ☐
- describe one benefit and one problem of globalisation for India. ☐

Ghana and its trade

See Core Book pages 106–107 and 110

In these activities you will:

- compare the imports and exports of Ghana and the UK
- consider some of the trade problems faced by Ghana.

1 Read page 106 and study Figure 26.

a) Complete the following sentences by underlining the correct answer.

Ghana's exports are largely agricultural products: they make up **33** per cent / **43** per cent / **53** per cent of all exports. However, in the UK, primary commodities form a **smaller** / **larger** / **similar** percentage of the exports.

Ghana's largest import is **agricultural** / **mining** / **manufactured** products and these make up about **55** per cent / **65** per cent / **75** per cent of all imports.

b) Complete the following summary in your own words.

Ghana trades with a lot of countries including ....

Ghana exports a lot of raw materials such as ....

....

Most of Ghana's imports are ....

2 Read pages 106–107. Match the following key words to their correct definitions.

a) imports **b)** exports **c)** quotas **d)** import duty

e) free trade **f)** protected trade **g)** subsidy

Key word	Definition
	Restrictions on the amount of particular goods that can be imported each year
	A tax placed on goods brought into a country to make them more expensive
	A payment made by the government to a key industry, such as farming, to keep costs low
	Goods purchased from another country
	Trade that takes place without any limits or control
	Goods sold to another country
	Limits to trade that are made to protect workers

3 Use Figure 27 on page 107 to complete the following flow chart in your own words.

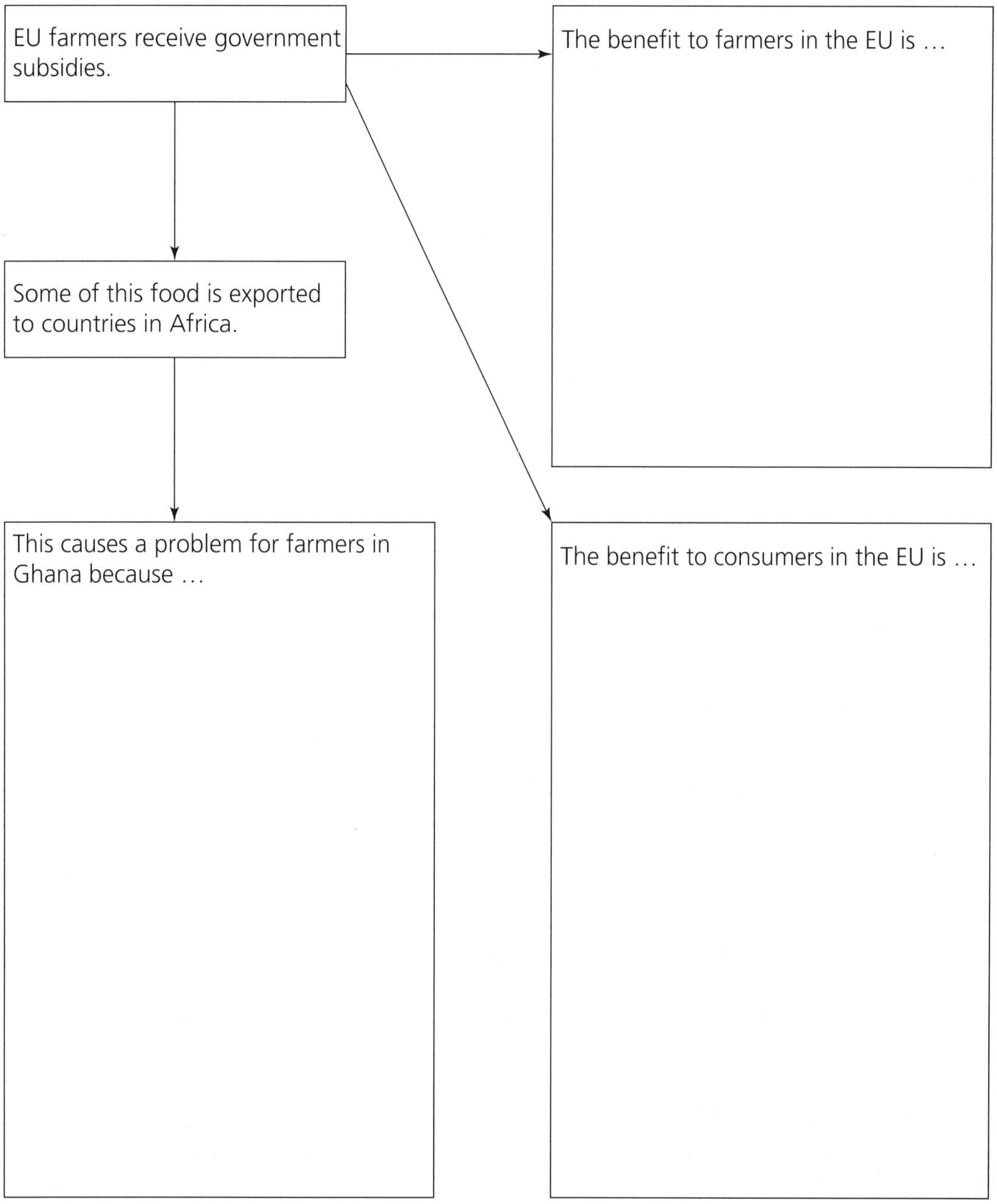

Figure A Impacts of EU subsidies

4 Read page 110.

a) Give three food products that are produced cheaply in the EU and exported to Ghana: ....

b) What is this practice called? ....

Review what you have learned

I can:

- describe Ghana's pattern of trade ☐
- give at least one problem for Ghana that is caused by its trade with other countries. ☐

Development

What are the regional patterns of development in Ghana?

See Core Book pages 116–119

In these activities you will:

- draw a sketch map of Ghana that shows why some regions are more highly developed than others
- practise the geographical technique of drawing a shaded (choropleth) map.

1 Study Figure 14 and the text on page 117. Underline the correct answers in the sentences below.

The regions which have most arable land (where food and cash crops are grown) are in the **south and west** / **east** / **north** of Ghana. This region has a climate that is **hot and wet** / **hot and dry**. In the far north of Ghana the climate is **cold** / **rainy** / **hot and dry**. The far north of the country has the **most** / **least** productive farm land.

2 Ghana suffers from a sharp north–south divide. One part of the country is more developed economically and socially than the other.

Read pages 116–119. Use information from these pages to complete the matching exercise below.

1. Accra and Central regions have Ghana's biggest cities and the country's best roads so …	… there are lots of small farms growing food crops like sorghum or cash crops like coffee. Average incomes are lower than in urban areas but better than in most other rural areas.
2. This region of Ghana has an international airport. The main beaches are here so …	… it is difficult to produce a lot of crops and farmers keep goats. Farmers' incomes are very low.
3. This region has fertile soil and lots of rainfall all through the year so …	… industry has grown faster here and new businesses find this part of Ghana more attractive.
4. The annual rainfall in this region is low and sometimes there is drought so	… there has been less investment in services such as electricity, clean water and schools so quality of life is poor.
5. This is a rural region. It has poor transport links and it is a long way from the larger cities of the south so …	… it has been easier to develop tourism in this part of the country and create jobs in the tourist industry.

3 Use the data in Figure 18 on page 119 to make a shaded map showing percentage of the population who have access to health services in different regions of Ghana.

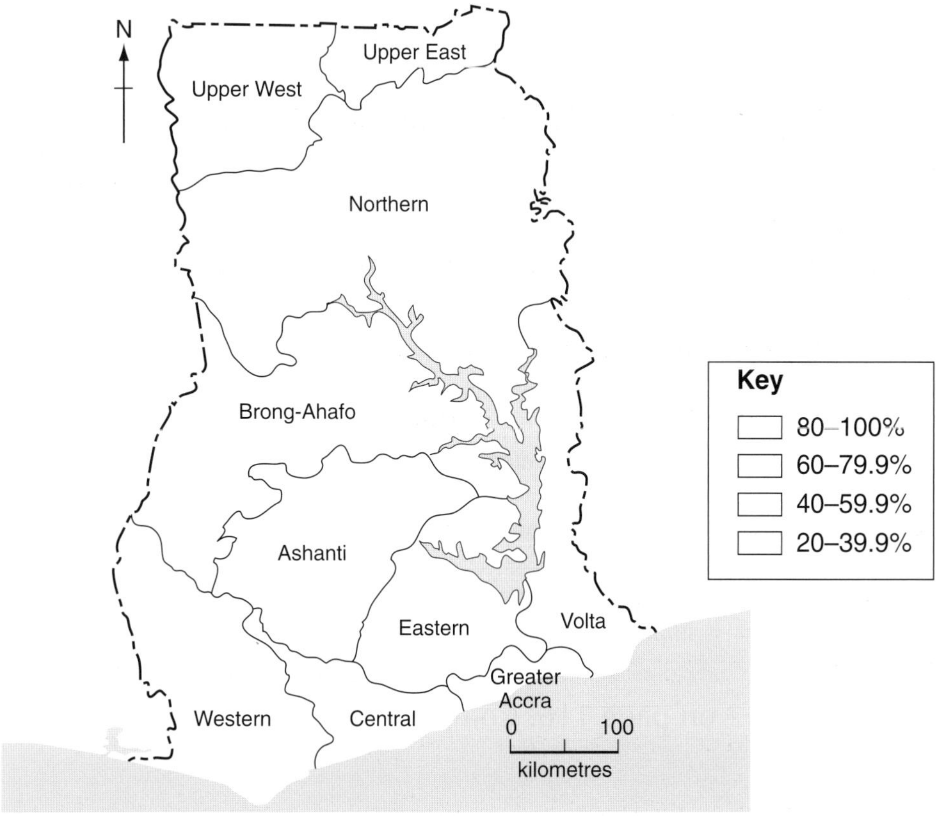

Figure A Outline map of Ghana's regions

a) Choose four colours for the key. Use the lightest colour for the smallest numbers.

b) Colour each region on Figure A using your key. For example, Western region has access of 46.9% to health services, so fits into the range 40–59.9%.

4 Compare your finished map to Figure 15 on page 118.

a) What are the similarities between the two maps?

b) What does this tell you about regional inequality in Ghana?

5 a) Draw a simple outline map of Ghana in your notebook.

b) Add labels for Atlantic Ocean, Lake Volta, Greater Accra, Ashanti and Northern region.

c) Shade the area of Ghana that has more than 60% permanent crops and arable land.

d) Add five numbered boxes to the map (one for each description that you completed in activity 2) and use a tie-line to make sure they are pointing to the correct region.

Review what you have learned

I can:

- draw a sketch map ☐
- draw a shaded (choropleth) map ☐
- give three reasons why some regions of Ghana have a higher level of development than other regions. ☐

How does aid help countries like Mali reach its MDG targets?

See Core Book pages 116, 122–124

In these activities you will:

- explain why female literacy is important to development
- develop a case study of an aid project in Mali.

1 Read page 124. Match the following key words to their correct definitions.

a) emergency aid **b)** development aid **c)** NGO **d)** MDG

Key word	Definition
	Organisations working in the voluntary sector
	Development targets set by the United Nations
	Aid such as food, shelter and medicine given after a disaster
	Aid such as training for farmers or teacher training/building schools to improve development over the long term

2 Study Figures 11 and 12 on page 116. Complete this description of Mali by underlining the correct answers.

Mali is in **West** / **East** / **Southern** Africa. The **northern** / **southern** part of Mali lies on the Tropic of Cancer. The maps show that Mali is **very small** / **landlocked** / **covered by desert**. The GNI (per person), which is a measure of income, in Mali was US$ **400–649** / **650–899**. The GNI in Niger was **the same** / **lower** / **higher**. The percentage of the population in Mali who are usually affected by undernourishment was **10–19** per cent / **20–29** per cent / **30–39** per cent. In Niger this percentage was **the same** / **lower** / **higher**.

3 Study page 122.

a) Why is it that some parents fail to send their daughters to school? Complete the following diagram in your own words:

Parents may keep their daughters at home because ...

→ ... the family is poor and ....

....

→ ... girls contribute to household income by ....

....

→ ... girls help around the house, for example ....

....

b) Give three reasons why the UN set MDGs that aim to give every child a primary education and improve educational opportunities for young women.

4 Study Figure 23 on page 122. How much progress was made between 1990 and 2004 in improving female education?

a) Complete the table below by calculating the differences. One answer has been completed for you:

	Sub-Saharan Africa	Middle East and North Africa	South Asia
Male literacy 1990	60	66	59
Female literacy 1990	40	39	34
difference	20		
Male literacy 2004	69	74	66
Female literacy 2004	53	52	42
difference			

b) Use the results of your table to identify which region:

i) has narrowed the gap most between male and female literacy: ....

ii) has made least progress in improving female literacy: ....

5 Read Figure 26 on page 124. It describes development aid aimed at improving the education of girls in part of Mali. Use the information to complete this summary of the project:

Who is providing the aid?	
Three facts about the community who are receiving the aid	1 2 3
Two facts that could be used as a measure of success of this project	1 2

Review what you have learned

I can:

- explain why it is important to improve female literacy ☐
- give an example of a development aid project aimed at improving female literacy. ☐

What links water, health and development?

See Core Book pages 118–119, 128

In these activities you will:

- practise the geographical technique of drawing divided bars
- describe how water supply can affect health and quality of life.

Water is needed for a healthy life. But many families in the world's poorest countries do not have enough clean water.

1 Read pages 118–119. Tick any of the following statements that are true:

Average household incomes in Greater Accra are almost twice that of incomes in rural savannah (Northern Ghana).	(✓)
Many people in Northern Ghana do not have access to safe drinking water.	
A much larger percentage of people in Northern Ghana have access to health services than in Greater Accra.	

Type of water supply		**In Accra**	**In savannah regions of Northern Ghana**	**In all of Ghana**
Pipe	Indoor	42.2	2.2	14.5
	Public standpipe	4.5	1.2	10.7
	Other	37.6	2.5	14.3
Pipe total		**84.3**	**5.9**	**39.5**
Well	Borehole or protected well	1.2	55.2	36.6
	Unprotected well	0	6.9	4.1
Well total		**1.2**	**62.1**	**40.7**
Natural e.g. streams and ponds		**0.2**	**31.9**	**15.7**
Other (includes water vendor)		**14.3**	**0.0**	**4.1**

Figure A Water supply in Ghana (%)

2 Study Figure A above. Use the data in Figure A to complete the two divided bar charts in Figure B for:

a) the savannah regions of Northern Ghana

b) all of Ghana.

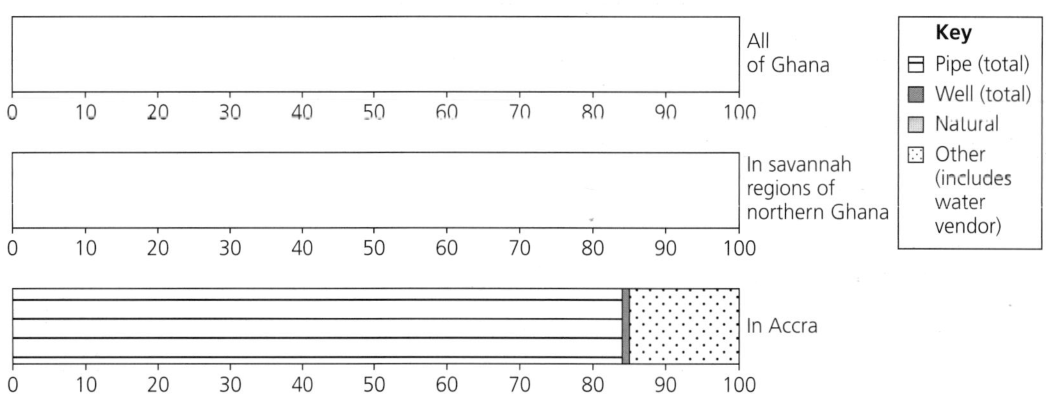

Figure B Water supply in Ghana (%)

3 Complete the following description of the graphs by underlining the correct answer and filling in the gaps using words from the box below:

wells	natural sources	pipes

In Accra most people get their water from .... whereas in the savannah regions most people get their water from .... In all Ghana about 16 per cent get their water from .... whereas in the savannah regions this percentage is **half / twice / three times the amount.**

4 A poor water supply can affect quality of life. Use the words and phrases in the box below to help complete each sentence to suggest how a poor water supply might affect a family.

expense	neck or back pain	wasted time	lack of privacy	sickness

a) Many women and children carry water long distances to their homes so … and …
b) People in shanty towns may have to buy water from a private seller in the street so …
c) People use streams for washing and collecting water so …
d) Some wells are unprotected from animal waste so …
e) Families who have no water supply have very small amounts of water for washing so …

5 How could the water supply be improved in:

a) Accra **b)** the savannah regions of northern Ghana?

6 Study Figure 33 on page 128.

a) Compare the progress made in each region between 1990 and 2002 with the progress needed between 2002 and 2015 to decide whether each region is likely to reach its target. Put a tick (✓) in the column that you predict.

Region	Likely to reach target	Unlikely to reach target
World average		
Arab states		
Latin America and Caribbean		
East Asia and the Pacific		
South Asia		
Sub-Saharan Africa		

Review what you have learned

I can:

- draw a divided bar graph ☐
- describe two ways that water supply can affect health. ☐

Case study: Water supply in South Africa

See Core Book pages 129–132

In these activities you will:

- show you know why it rains in Lesotho more than in South Africa
- describe advantages and disadvantages of the LHWP.

The pattern of rainfall in South Africa is very uneven. Most rain falls in the mountains in the east. This region includes Lesotho, a small country that is entirely surrounded by the much larger South Africa.

1 a) Study Figure A below.

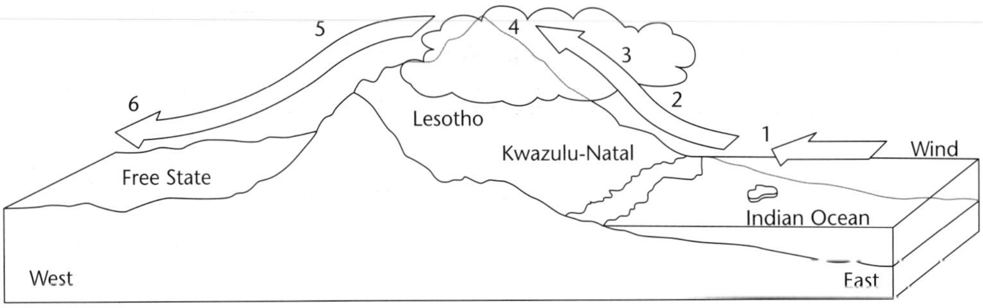

Figure A Diagram of rainfall patterns

b) Use words from the box to complete each label (1–6) for the diagram.

evaporates / condenses	**leeward / windward**	**descends / ascends**
prevailing / shadow	**leeward/windward**	

1. Water vapour .... from the warm Indian Ocean. This moisture is picked up by the prevailing wind.
2. Air rises up the .... slope of the mountains in KwaZulu-Natal.
3. As the air .... it cools. The water vapour .... into small water droplets.
4. Water droplets bump into one another. Clouds form and it rains in Lesotho.
5. The air crosses the mountains. The temperature rises as the air ....
6. The air on the .... side of the mountains is now warm and dry. This is called the rain ....

The Lesotho Highlands Water Project (LHWP)

The government of Lesotho agreed to sell some of its water to South Africa. Dams have been built to collect the water. Tunnels take the water through the mountains into South Africa. It then flows down rivers towards Johannesburg where it is used.

2 What are the advantages and disadvantages of the LHWP? Read the statements opposite (based on evidence on pages 129–130).

a) Sort the statements by colouring advantages in green and disadvantages in red.

b) Decide whether each statement describes an advantage/disadvantage for Lesotho or South Africa. Add L (Lesotho) or SA (South Africa) to the final column.

Advantages (green) and disadvantages (red) of the LHWP	L or SA?
Lots of construction jobs were created when the dams and tunnels were built.	
The percentage of people with a safe water supply in this country has increased.	
In the future, illness caused by drinking dirty water should be very rare.	
People in Johannesburg paid higher water bills to help pay for the LHWP.	
This country receives income from selling water to the other. The LHWP earns 75% of the country's income.	
Income from the LHWP could help this country develop its own water management schemes in the future.	
Local people have not had any of the water for their own homes.	
Farmland was flooded when the dams were built.	
Losing so much farmland may make it difficult for this country to grow enough food in the future.	
If the project is completed it will affect the quality of river water. Wildlife may be affected by lower oxygen levels and colder water temperature.	
The dams provide cheap hydro-electric power (HEP).	
The government of this country had to borrow US$8 billion to pay for the dams and tunnels. This money will take many years to repay.	
Many people lost their homes when their land was flooded by the new dams.	
Improved roads were constructed to gain access to the dam sites.	
Around 20,000 people moved into informal settlements to work on the dams. Social problems such as HIV and alcoholism became an issue.	

3 Look at your completed table above. Overall, do you think the LHWP was a good water project for **both** countries? Explain your decision.

4 The LHWP is a large-scale water management project. It was very expensive. Use pages 131–132 to investigate small-scale, cheap ways of managing water. Make a poster that describes how these projects work. Explain their advantages.

Review what you have learned

I can:

- explain why it rains in Lesotho ☐
- give one advantage and one disadvantage of the LHWP for Lesotho. ☐

Our Changing Coastline

Erosion at the coast

See Option Topics Book pages 1–5

In these activities you will:

- show you know how cliffs are eroded.

1 a) Study Figure 5 on page 3. It shows how cliffs made of resistant (hard) rocks retreat when they are eroded.

b) Match the labels for Figure A to the correct point in the diagram by adding arrows.

c) Add a number to each label to suggest the best sequence in which to read them.

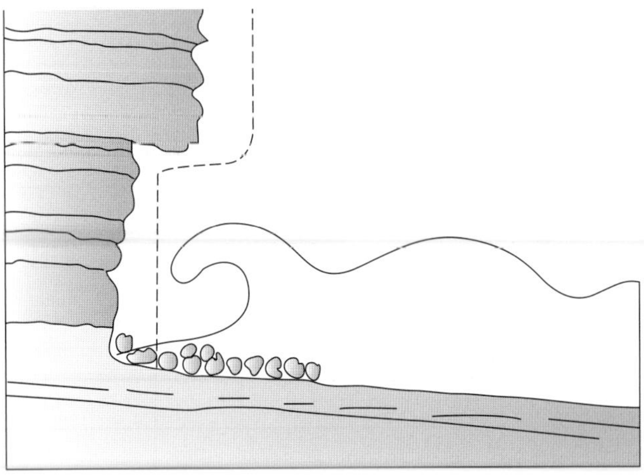

The cliff retreats leaving a wave-cut platform behind.

The cliff forms an overhang over the eroded notch.

Waves use pebbles from the beach to erode a notch at the foot of the cliff.

Figure A Cliff retreat

2 Study Figure 6 on page 5 and Figure B below. Write labels that describe the landforms and processes in boxes A, B, C and D. Use words from the box below to help you.

limestone	**clay**	**quickly eroded**	**resistant to erosion**
stumps	**bay**	**slumping**	**landslips**
beach	**swash**	**backwash**	

Figure B A cove on the Jurassic coastline of Dorset

3 Study Figure 1 on page 1. The table below describes the features and processes occurring at A, B, C and D on the photograph.

a) Complete each annotation in the table by adding words from the box below. You will need to use some words more than once.

b) Match each of the annotations to one of the features at A, B, C or D by adding the correct letter to the first column in the table.

spit	**erosion**	**backwash**	**stacks**	**abrasion**
longshore drift	**swash**	**hydraulic action**	**weathering**	

Letter	Label
	Waves move sand up the beach in a process known as .... Some sand is dragged back into the sea by the ....
	Black volcanic sand has moved along the beach (towards the camera) by the process of .... Over many years this process has created a long ....
	Crashing waves compress air into cracks. This process is known as .... The cliffs are also smashed by rocks carried by the waves. This process is known as .... Together these processes have caused .... of the cliffs and have created a group of ....
	This stack was created by .... when it was once surrounded by water. Over the years the .... has grown and the rocks are now protected from the waves by the beach. The rocks at the top of this feature will be affected by .... from frost and rainwater.

Review what you have learned

I can:

- use, with confidence, four key words that describe the processes of coastal erosion ☐
- add labels to a photograph to describe its coastal features. ☐

Beach processes and the formation of spits

See Option Topics Book pages 6–7

In these activities you will:

- show you know how the process of longshore drift works.

1 Study Figures 9, 10 and 11 on pages 6–7. Use words from the box below to describe what processes may be happening at each label, A, B and C.

beach	**sand dunes**	**wave-cut platform**	**spit**	**hydraulic action**	**abrasion**
swash	**backwash**	**attrition**	**transportation**	**deposition**	

A is a ....

It has been formed by ....

....

At **B** there is a long ....

It has been formed by ....

....

Over a long period of time this feature will change by ....

....

At **C** there are ....

These features are formed when sediment is swept up the beach and ....

....

2 Study Figure 8 on page 6 and underline the correct answers in the passage below.

The movement of sand along a beach is a process called **erosion** / **longshore drift** / **deposition**. The waves are pushed up the beach at an angle by the **backwash** / **wind** / **tide**. This is called the swash and it deposits beach material like pebbles and sand on the beach.

Some of the beach material is picked up by the waves and carried back off the beach by gravity. This is called **backwash** / **deposition** / **swash**. The swash and backwash move sediment along the beach in a zig-zag pattern.

3 The process of longshore drift has created a depositional landform known as a spit at Borth. Study Figure 11 on page 7. Use information from this diagram to complete the labels in Figure A.

Waves use abrasion and hydraulic action to .... the cliffs between Aberystwyth and Borth

Sand is transported down the River .... and carried out into the sea

The cliffs retreat leaving a flat rocky structure (seen at A in Figure 9) known as a ....

Pebbles from the cliffs are transported in a .... direction along the coast. This process is called ....

Over many hundreds of years this process builds a feature (seen in B in Figure 9) known as a ....

This sand is .... in a huge pile on the sea bed known as an ....

Wave swash transports sand from the bar onto the ....

The wind blows the sand up the beach. It is .... on the leeward side of each ....

Figure A The processes affecting the coastline at Borth

4 Show you understand coastal landforms and processes.

a) In each list, circle the odd one out.

					Common theme
1.	Swash	Backwash	Longshore drift	Spit	
2.	Arch	Spit	Stack	Cave	
3.	Abrasion	Deposition	Hydraulic action	Corrosion	

b) For each list, state what the three remaining words have in common. Write your answer in the final column of the table.

Review what you have learned

I can:

- describe how sediment is moved along the coast by longshore drift. ☐

Case study: Coastal flooding

See Option Topics Book pages 11–12

In these activities you will:

- describe the causes and effects of a coastal flood
- practise drawing and labelling a sketch map.

1 Study Figure 18 on page 11. It shows a coastal landscape known as salt marsh. Choose appropriate words from the box to write a short description of this landscape.

flat	**hilly**	**tributaries**	**creeks**	**meandering**
drainage	**natural**	**boggy**	**low-lying**	

....

....

....

....

2 In 1953 the east coast of England was affected by terrible coastal floods.

Study Figure 19 and Figure 20 on page 11. Use words from the box to complete the description of the causes of this flood event.

low	**high**	**tide**	**flood**	**storm**
wind	**physical**	**human**	**natural**	

The .... factors that caused the floods were a combination of .... tides and a massive .... surge causing sea levels to rise. A surge occurs when air pressure is very ....

The flood was made worse at Jaywick because people had built homes on the .... plain. This could be described as a .... factor that contributed to the disaster.

3 The 1953 floods in numbers: summarise the effects of the 1953 floods by adding a number to each box.

Miles of coastline flooded		Acres of land flooded in Essex	
Number of people evacuated		Number of people who died in Essex	
Number of people who died in Jaywick		Number of people who died overall	

4 Study Figures 22 and 23 on page 12 and Figure A below. Use the grid lines on Figure 23 to draw a sketch map of this coastal area on Figure A below. You should include:

- the coastline
- St Osyth Creek
- the settlements at Point Clear, Seawick and Jaywick.

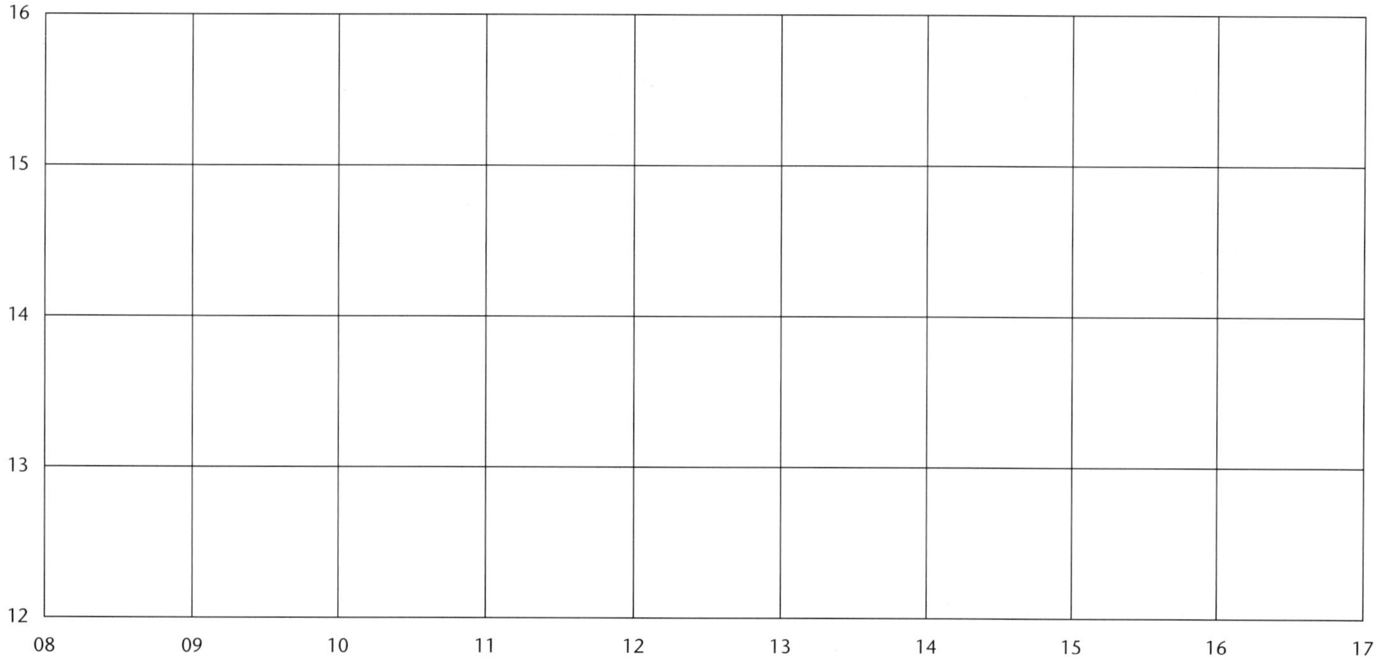

Figure A Sketch map of the coastal area west of Jaywick

5 The east coast of England was hit by severe coastal floods on 5 and 6 December 2013. These were the worst floods since those that flooded Jaywick in 1953.

a) Use the internet to research the floods of 5 and 6 December 2013.

b) Make notes on the causes and effects of these floods.

c) How did the 2013 floods compare with the 1953 floods?

6 Should Jaywick be protected from future floods? Read each of the points of view in the table below. Decide whether each is an argument to hold the line or retreat the line.

	Hold the line (✓)	**Retreat the line (✓)**
St Osyth Marsh is low grade agricultural land. It would make a natural barrier if the sea was allowed to flood it at high tide.		
Some families have lived in Jaywick for generations. They deserve to be protected from flooding.		
Coastal defences are very expensive and sea levels will rise in the future. Money should be spent on hard engineering at larger coastal towns in Essex such as Clacton.		

Review what you have learned

I can:

- describe the physical causes of a coastal flood ☐
- describe three effects of coastal flooding ☐
- draw and label a sketch map. ☐

Coastal management

See Option Topics Book pages 6, 13, 14, 19, 21

In these activities you will:

- describe different ways that coastlines may be managed
- consider the advantages and disadvantages of one coastal management strategy.

1 a) Study each of the **four** management strategies listed in the table below by reading the page numbers (given in the first column of the table).

b) Describe how the strategy works and give a located example to complete the table.

✎ **c)** Choose **one** of these strategies and describe its advantages and disadvantages.

Coastal management strategy	How it works	Location
Wooden groynes Figure 8 page 6		
Artificial rock reef Figure 24 page 13		
'Sandtainer' reefs Figure 26 page 14		
Managed realignment Figure 40 page 21		

2 Study the information on page 19 which shows how Australia is trying to manage the impacts of coastal erosion. Use the information to complete the newspaper article opposite.

Coastal change threatens Australia's coastal lifestyle, warns report
Thousands of miles of Australian coastline are under threat from rising sea levels.

Coastal erosion has social impacts
People are worried because:

....

....

....

....

Three important facts:
-
-
-

Examples of **three** places that are being affected:
-
-
-

Coastal erosion has economic impacts
The damage to the coastline will cause economic problems such as:

....

....

....

....

Different stakeholders have different views
Some groups of people are in favour of building more sea defences because:

....

....

....

....

Other groups of people are against building more sea defences because:

....

....

....

....

Review what you have learned

I can:
- describe three different ways to manage the coast ☐
- identify the advantages and disadvantages of one coastal management strategy ☐
- describe the views of two different stakeholder groups. ☐

Weather and Climate

Weather or climate?

See Option Topics Book pages 23–27, 29

In these activities you will:

- identify the differences between weather and climate
- interpret synoptic charts (weather maps)
- describe some of the factors that make our weather and climate so varied.

1 Study page 23 and the key words below. Tick (✓) the correct definition for each term.

a) What is **weather**?

Temperature readings taken over a period of time ☐
Day-to-day conditions of the atmosphere, e.g. temperature, rain, hail, sunshine ☐
The amount of rainfall in one day ☐

b) What is **climate**?

Average weather readings taken over a period of around 30 years ☐
Average rainfall of a given place ☐
Average temperature of a given place ☐

2 Identify whether each of these statements describes weather or climate.

	Weather or climate?
It was a very hot day in London.	
January is usually a good time to head to Australia to catch some winter sun.	
The monsoon season is from May to September in South East Asia.	
A terrible storm in Rochdale brought strong winds and over 100 mm of rain.	
Fog on the motorway reduced visibility to less than 100 metres.	
Egypt is usually very hot in August.	

3 Read page 26 and study Figure 5. Complete the table below to describe the weather brought to the British Isles by four different air masses.

Type of air mass	**Direction air mass comes from**	**Weather it brings**
Polar maritime		
Tropical maritime		
Polar continental		
Tropical continental		

4 The type of air masses over the UK is one factor that affects our weather and climate. Read pages 26 and 27, and study Figures 6, 7 and 8. Use this information to complete the paragraph below by underlining the correct answers and filling in the gaps.

The British Isles has a **10°** / **20°** difference in latitude between the north and south of the country. This helps explain why the average maximum temperature in Oxford is **2.6** / **3.6** / **4.6**°C warmer than in Kirkwall. In winter the west coast of Britain is **warmer** / **colder** than other places at the same latitude. For example, Plymouth and Prague are both at **45** / **50** / **55** °N latitude but in January Plymouth is 1 °C warmer than Prague. This is due to the .... which is a warm current of water. The sea is able to **retain** / **release** its heat in winter and cools down very slowly. That is why the **south-west** / **south-east** of Britain is warmer in winter than the north-east.

5 Study Figure 9 on page 27. You may find that using an atlas or Google Earth is helpful for this activity.

a) Decide whether each statement below is describing what is happening at A or B.

A or B	Statement
	Warm water
	Cold water
	Ocean currents moving away from the coastline of Florida
	The ocean helps to keep New York colder than the UK in winter
	Ocean currents moving southwards from colder Arctic regions
	The source of warmth that travels across the Atlantic in the Gulf Stream
	Tropical sunshine is warming the water
	Cold water warms up very slowly as it moves south

✎ **b)** Use the ideas in the table above to write your own sentence that explains how the west coast of western Europe is kept warm by the Gulf Stream.

Review what you have learned

I can:

- explain the difference between weather and climate ☐
- read a synoptic chart ☐
- give three reasons why Britain has such varied weather. ☐

Continental and maritime climates

See Option Topics Book pages 26–27, 49

In these activities you will:

- investigate winter temperature patterns in Europe
- discover the influence of the Atlantic Ocean on Europe's climate.

The ocean does not get as cold as the earth during winter. The Atlantic Ocean acts a bit like a hot water bottle during the winter and warms up the surrounding areas! It gives western Europe a winter that is wet and warm. This is a **maritime** climate. Eastern Europe, which is further from the ocean, is much colder in winter. It has a **continental** climate.

1 a) Complete this prediction by underlining the correct words.

I expect that winter temperatures in western Europe will be **warmer** / **the same** / **colder** than in eastern Europe. I also expect temperatures to get **warmer** / **colder** as you travel from the Mediterranean towards the Arctic. The coldest part of Europe in winter will be the **south-west** / **east** / **north-east**. The warmest part will be the **north-west** / **south-west** / **south-east**.

b) Plot the temperatures in Figure A on the right onto Figure B opposite. Use an atlas or the internet to find each location.

c) Add colour to your map by shading each shape made by the lines of latitude and longitude. Use 'warm' colours like orange for the temperatures above zero. Use 'cold' colours like shades of blue for temperatures below zero.

d) Add a key to your map.

e) Were the predictions you made at the start of this enquiry correct? **Yes** / **No**

✎ **f)** Compare the temperatures in Iceland to those in Finland and Russia at the same latitude. Explain this difference in temperature.

Location	Temp °C
Lisbon	8
Rome	5
Bordeaux	2
Reykjavik	–2
Berlin	–3
Ankara	–4
Warsaw	–6
Oslo	–7
Bucharest	–7
Helsinki	–9
Kiev	–10
St Petersburg	–13

Figure A Average minimum January temperature °C

2 Why is it colder in Arctic Europe than in the Mediterranean? Study Figure 9 on page 49. Underline the correct words in the following passage.

In the northernmost parts of Europe, the Sun's energy hits the ground at a **low angle** / **right angle**, so winter temperatures are very **low** / **high** and the summer season is **short** / **long**. The Sun is higher overhead in southern Europe, so winter temperatures here are **colder** / **warmer** and summers are **shorter** / **longer** than in the Arctic.

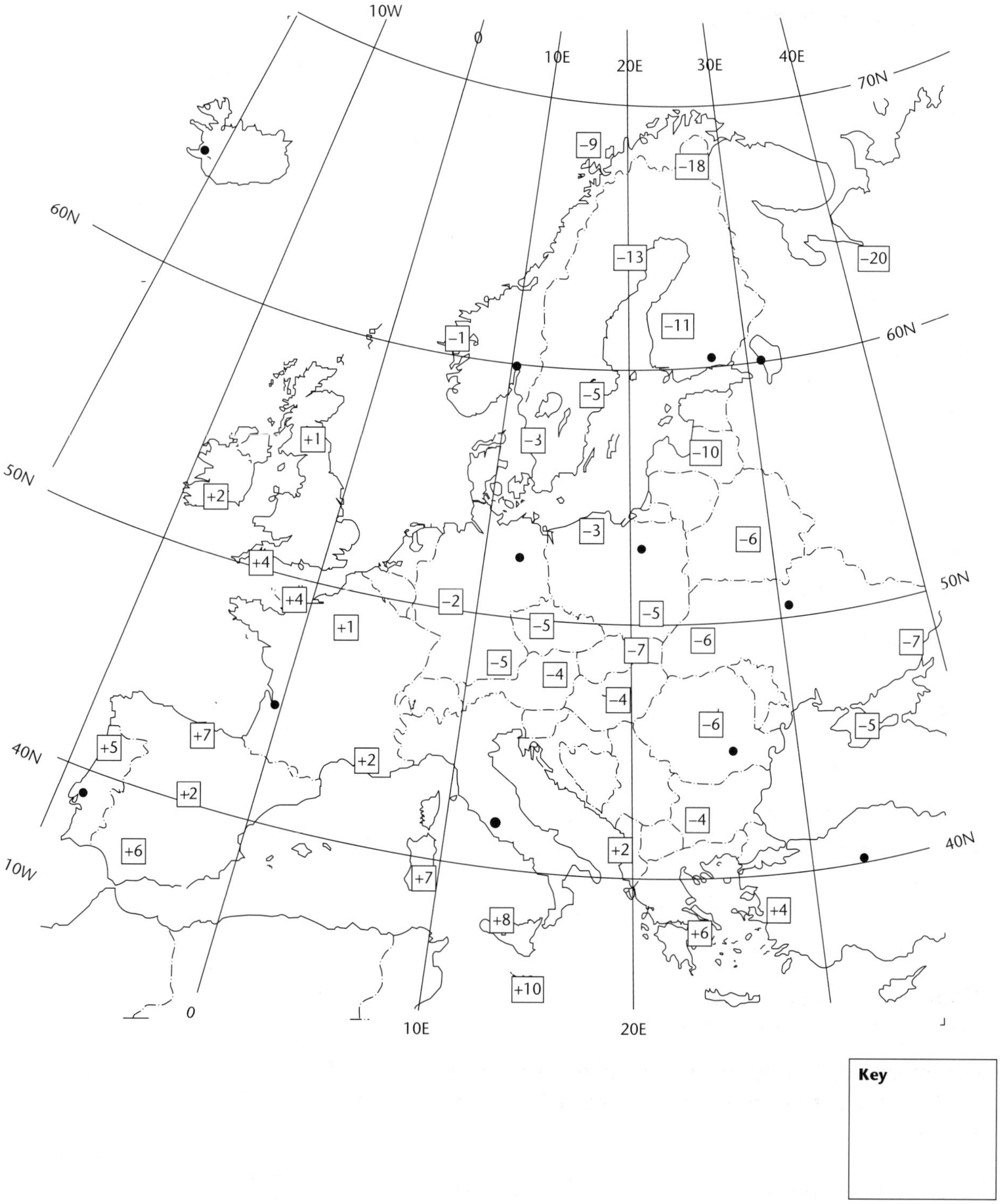

Figure B Average minimum January temperatures across Europe °C

Review what you have learned

I can:

- use an atlas to find the location of cities ☐
- describe the pattern of winter temperatures across Europe ☐
- explain why western Europe has a warmer climate than eastern Europe in winter. ☐

Low pressure systems

See Option Topics Book pages 32–33 and 35–36

In these activities you will:

- practise reading from a weather map (synoptic chart)
- develop a case study of a tropical cyclone.

1 Study Figure 18 on page 32. It shows low pressure (a depression) over Iceland. Show you understand this weather map by adding the following details to Figure A.

a) Colour the wedge of warm air in red and the area of cold air in blue.

b) Label the 1000mb and 1016mb isobars.

c) Add the letter L to the centre of the low pressure.

d) Label one place on the map where you expect wind speeds are strong.

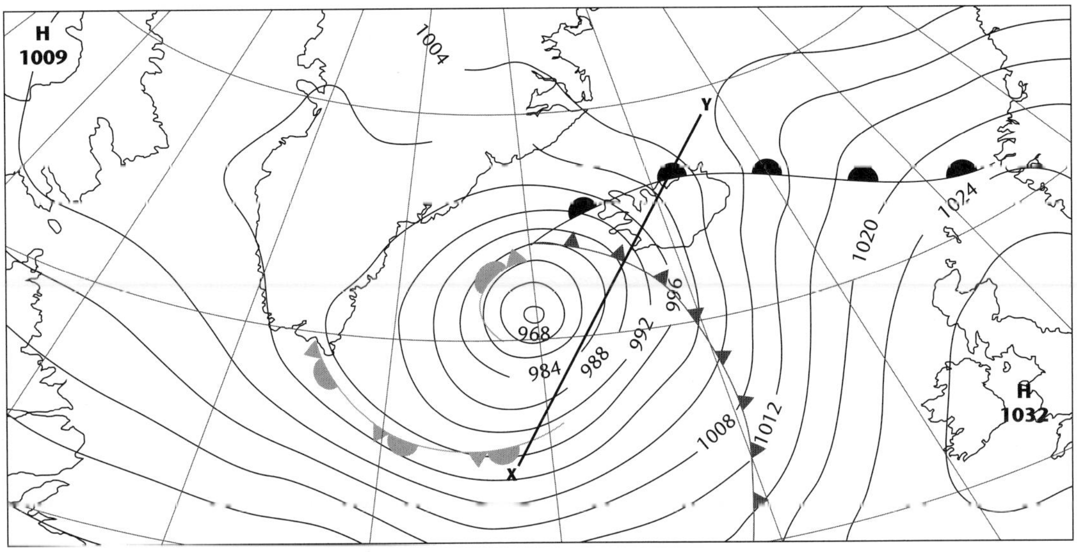

Figure A Weather map (synoptic chart) showing a low pressure system

2 Complete Figure B, which is a cross-section (X–Y) through the depression in Figure A.

a) Colour the warm sector in red and the cold sector in blue.

b) Add labels a)–h) to Figure B from the box below:

a) cold front	**b) warm front**	**c) stratus clouds**	**d) cirrus clouds**
e) cumulus clouds	**f) cumulo-nimbus clouds**	**g) heavy rain**	**h) rain**

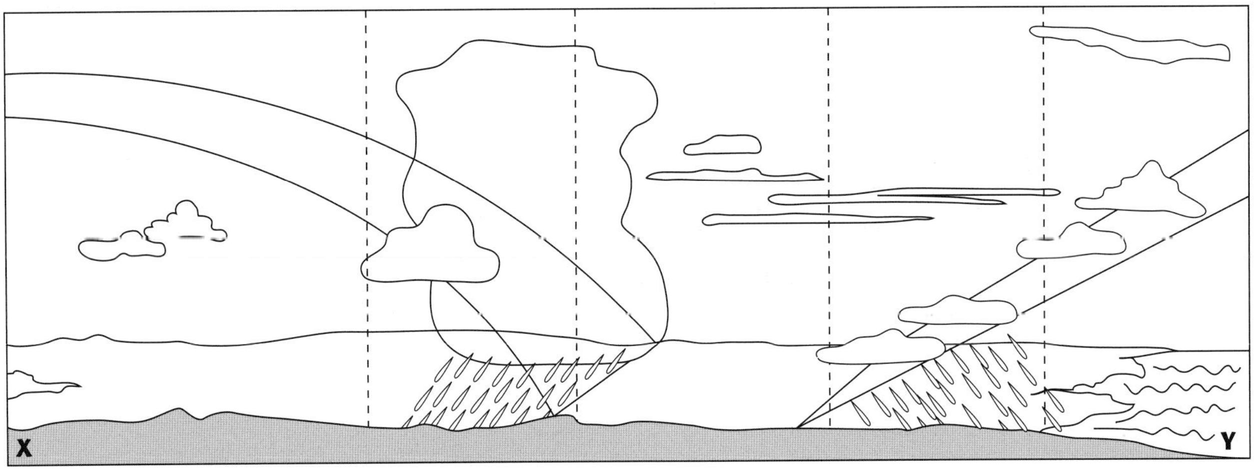

Figure B Cross-section X–Y through the depression on Figure A

Case study: Cyclone Nargis

3 Study Figure 24 on page 35. Describe the track of Cyclone Nargis by underlining the correct answers in the following passage.

Cyclone Nargis started as a **tropical depression** / **tropical storm** / **hurricane** in the Indian Ocean on 29 April 2008. It travelled in a **south-westerly** / **north-westerly** / **south-easterly** direction until 30 April. It then changed direction and moved towards **India** / **Burma (Myanmar)** / **Sri Lanka** as a category 1 hurricane. On 2 May it had become a category **2** / **4** / **5** hurricane and made landfall on the coastline of Burma (Myanmar). It travelled across the country in a **north-easterly** / **south-westerly** direction and eventually ended on 4 May.

4 Cyclone Nargis was very destructive. The table below describes some of its features and some of its effects.

a) Colour the features of the storm in green.

b) Colour the effects of the storm in red.

Cyclone Nargis was formed over water with a temperature of at least 26 °C.	An estimated 300,000 people were killed.	The wind speed reached 215 km/hr.	Three-quarters of health centres in the region affected by the cyclone were damaged.
The low air pressure caused the ocean to bulge upwards in a storm surge.	The storm surge flooded the coastal areas destroying housing and rice crops.	The strong winds created very large waves which were 7.6 m high on top of the storm surge.	About 800,000 homes were damaged.
Many people had to live with family members because they had been displaced.	There were food shortages because crops had been destroyed by the flooding.	The hurricane brought very heavy rainfall to Burma (Myanmar).	People got diarrhoea from drinking water which had been contaminated with sewage.

Review what you have learned

I can:

- identify features on a weather map ☐
- describe three facts about the track and strength of Cyclone Nargis ☐
- describe three effects of Cyclone Nargis. ☐

Case study: Drought in Barcelona

See Option Topics Book pages 31, 33, 41

In these activities you will:

- describe the weather features of a high pressure system
- develop a case study of drought in Barcelona.

1 High pressure (an anticyclone) creates different features in the weather during the summer and the winter. Read page 31 and study Figure 21 on page 33. Match up the features with the correct statements to describe anticyclones by drawing a line between them.

In an anticyclone, air pressure is …	… calm and moves at low speeds.
In an anticyclone, the air is …	… cold and dry with clear skies and frost at night.
In an anticyclone, the wind is …	… high, usually above 1020 mb (millibars).
In an anticyclone, wind moves in …	… a clockwise direction.
In a winter anticyclone, the weather is …	… sinking and becoming warmer.
In a summer anticyclone, the weather is …	… warm and dry. Heatwaves are common.

2 Study Figure 34 on page 41. Complete the sentences below using words from the box to describe the location of Barcelona.

Tarragona	**Marseille**	**Rhone**	**France**	**Spain**	**Zaragoza**
Mediterranean	**Llobregat**	**Segre**	**Catalonia**	**Aragon**	**Ebro**

Barcelona is a city in the north-east of the country of .... in the region of .... It is located on the mouth of the River .... It is located on the coastline of the Mediterranean Sea and is north-east of the city of ....

3 In 2007–2008 Barcelona suffered from a severe water shortage following a period of high pressure weather. Read page 41 and complete the table below.

	Evidence / example
Water use was restricted in public places.	
People who broke drought order rules were fined.	
The city was desperate and had to bring water from outside the area into the city's port.	

4 Study Figure 35 on page 41. It examines the different viewpoints surrounding the water shortage situation in Barcelona. Use this information to help you complete the following sentences.

The householder thinks that one reason for the water shortage is ....

because ....

....

....

....

The climate expert is against the idea of ....

because ....

....

....

The protestor in France is against the idea of ....

....

because ....

....

....

....

✎ **5** What do you think the authorities in Barcelona should do to reduce the risk of a severe water shortage in the future? Explain your ideas giving three clear points based on information on page 41.

Review what you have learned

I can:

- describe three weather features of a high pressure system ☐
- describe the location of my case study (Barcelona) using a compass and scale line ☐
- describe three ways that people in Barcelona were affected by drought. ☐

Living Things

Arctic ecosystems

See Option Topics Book pages 45, 50

In these activities you will:

- show you know how living and non-living things are linked in the Arctic ecosystem
- explain the reasons for slow plant growth in the Arctic.

1 Study Figure A on page 45 and the labels below. It shows the ways that living and non-living things are linked together in an Arctic ecosystem. Write the number of each label into the correct box in Figure A.

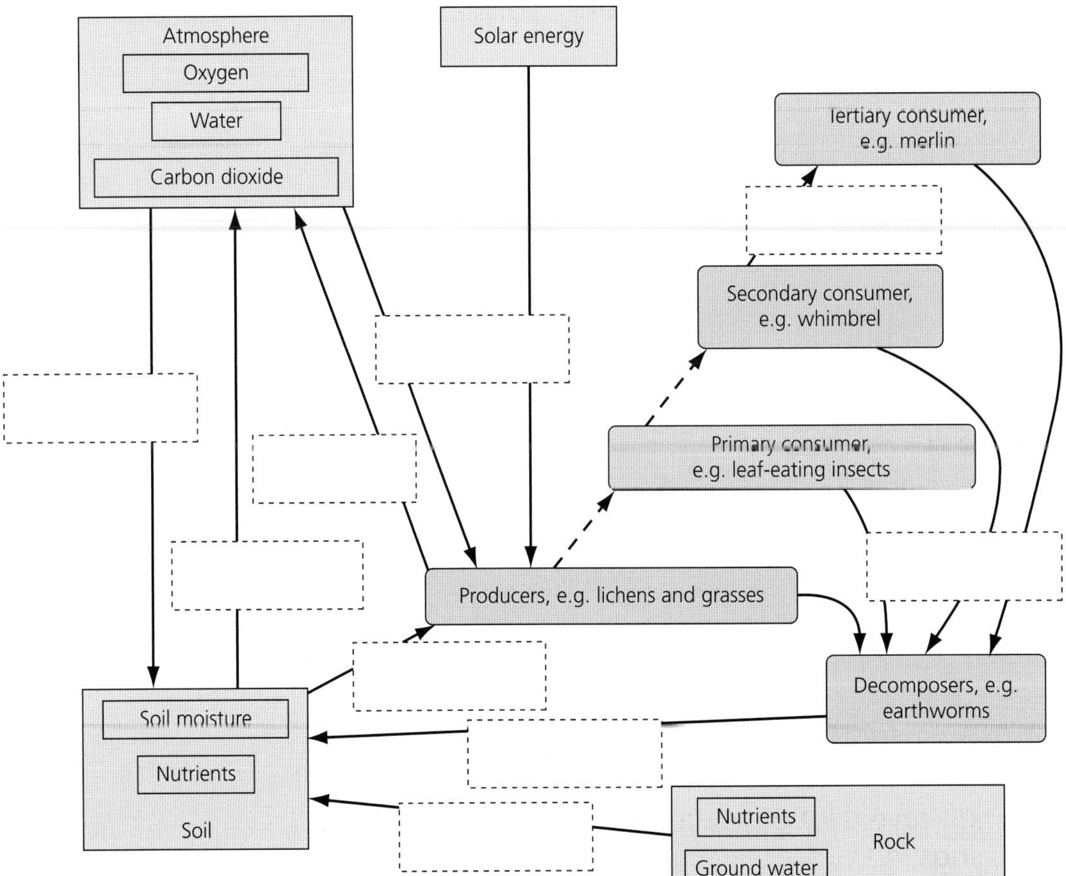

Figure A The tundra ecosystem of Iceland

1 Precipitation	**2** Evaporation
3 Weathering of rocks releases nutrients into the soil	**4** Plant roots take up water and nutrients from the soil
5 Decomposition breaks down dead matter and the nutrients are released into the soil	**6** The loss of water from pores in the leaves of plants (known as transpiration)
7 Dead plant and animal matter	**8** Carbon dioxide is taken into leaves and converted to starch by photosynthesis
9 Flow of energy through the food chain	

Labels for Figure A

2 Plants in Arctic ecosystems grow very slowly. Complete the following sentences to explain why they are so slow growing. Use words from the box below and the glossary to help.

nutrients	**leaching**	**transpiration**	**decomposition**
photosynthesise	**leaf litter**	**slowly**	**quickly**

Winter days are short and dark so plants are unable to .... In many months, water is frozen in the soil so plants have small leaves to reduce water lost through .... Bacteria in the soil reproduce very slowly in the cold weather so .... of dead plant matter is very slow. So nutrients are transferred from .... into the soil very .... The chemical reactions that release .... from rocks into the soil are slow because of the cold temperatures.

3 Study Figure B. It shows a food web in the Arctic ecosystem of northern Norway.

a) Add arrows to the diagram to show that:

i) reindeer eat lichen and moss

ii) Arctic fox eat lemming and young reindeer.

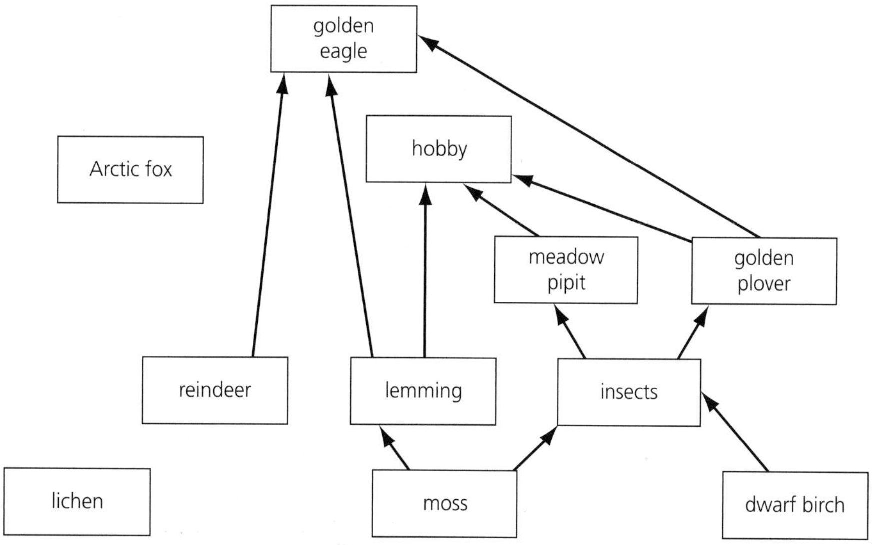

Figure B Food web in Arctic Norway

✎ **b)** Lemmings have a lot of young and breed quickly. Suggest what will happen to other living things in the food web if:

i) the population of lemmings gets very large

ii) the lemming population grows so large that it runs out of moss (which grows very slowly).

Review what you have learned

I can:

- identify three living and three non-living parts of an Arctic ecosystem and describe how they are linked ☐
- give three reasons for the slow growth of plants in the Arctic ☐
- describe an Arctic food web. ☐

Tropical ecosystems

See Option Topics Book pages 51, 65

In these activities you will:

- show you understand how water is recycled in the rainforest
- explain the reasons for rapid plant growth in the rainforest.

1 Study Figure 40 on page 65 and Figure A below. These diagrams show how water is recycled through the tropical rainforest.

a) Match the labels to the diagram by writing the correct number in each box.

b) Colour the water stores in blue. Colour the water flows (the processes that move water from one store to another) in red.

c) Make a key for your diagram.

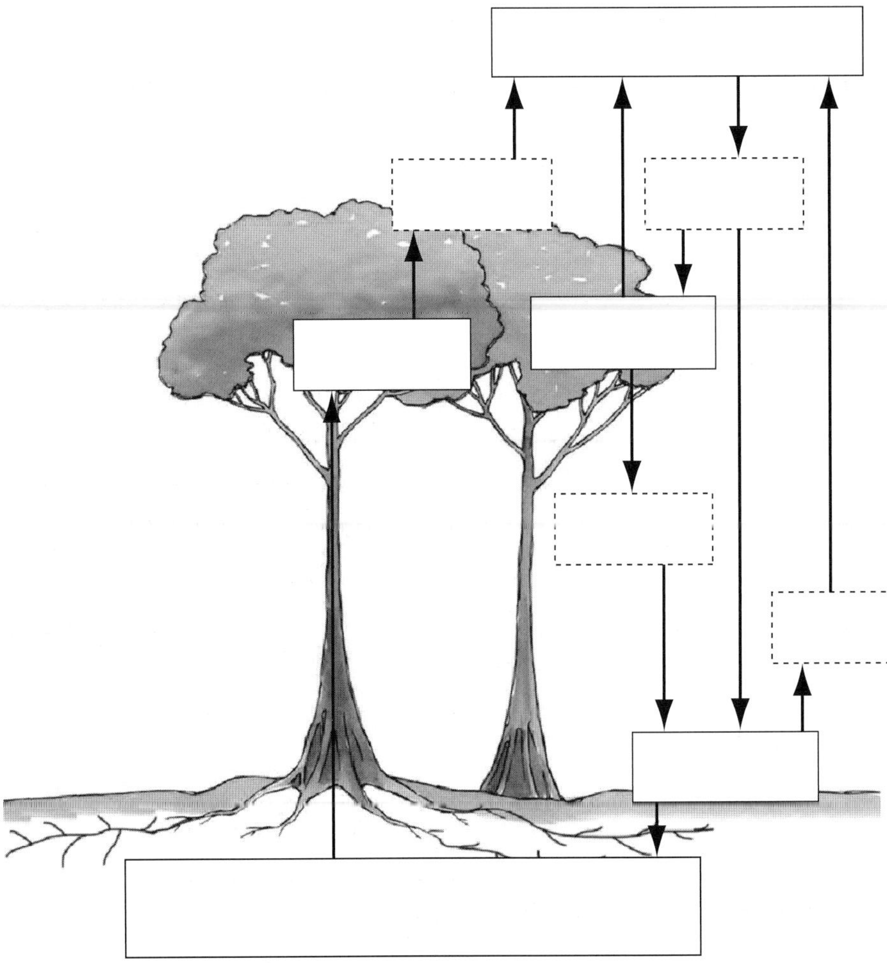

Figure A Water stores and flows in a tropical rainforest ecosystem

1 Evaporation	**2** Evapotranspiration	**3** Moisture in atmosphere
4 Water in the leaves and trunk of the tree	**5** Moisture in the soil	**6** Raindrops intercepted by the canopy
7 Stem flow and drip flow	**8** Precipitation	**9** Water in puddles and rivers (surface stores)

Labels for Figure A

2 Plants in the tropical rainforest grow very quickly. Complete the following sentences to explain why they grow so quickly. Use words from the box below and Figure 14 on page 51 to help you.

nutrients	**leaching**	**transpiration**	**decomposition**	**rapid**	**slow**
weathering	**photosynthesise**	**soil**	**rocks**	**slowly**	**quickly**

The winter months are warm and sunny so plants are able to .... all through the year.

Bacteria in the soil reproduce very .... in the hot weather so .... of dead plant matter is very rapid. So .... are transferred from leaf litter into the .... very quickly.

The chemical reactions that release nutrients from .... into the soil are .... because of the heat. This process is known as .... If trees are cut down, heavy rainfall can wash nutrients out of the soil. This process is known as ....

3 a) In each row, circle the odd one out.

					Common theme
1.	Mangrove forest	Cloud forest	Tropical rainforest	Boreal forest	
2.	Lion	Orangutan	Howler monkey	Boa constrictor	
3.	Canopy	Nutrient	Buttress root	Drip tip	
4.	Tobago	Brazil	Ghana	Iceland	
5.	Evaporation	Interception	Photosynthesis	Stem flow	

b) For each row, state what the three remaining words have in common. Write your answer in the final column of the table.

Review what you have learned

I can:

- identify three stores and three flows of rain water through the tropical rainforest ☐
- give three reasons for the rapid growth of plants in the tropics. ☐

Effects of rainforest destruction

See Option Topics Book pages 50, 54

In these activities you will:

- consider the effects of cutting down or burning large areas of forest.

1 Study Figure 12 on page 50. Complete the following sentence. Use words from the box in your answer.

canopy	**interception**	**raindrops**	**slope**

If the forest is clear felled all the trees are cut or burned. The soil is at risk of erosion

because ....

....

2 Study Figure 20 on page 54 and Figure A below. It shows how rivers can be damaged by logging.

Figure A Logging next to a river in Borneo

a) Complete each of the labels below. The first one has been done for you.

b) Then match the labels to locations A, B, C and D on Figure A.

Letter	Label
	Large machinery may leak oil / diesel into the river so ... *the drinking water of people living downstream is polluted.*
	The remaining trees are an ecological island so ...
	Soil has been washed into the river so ...
	Trees next to the river have been felled and the canopy has gone so ...

3 If very large areas of rainforest are cleared, this can affect the water cycle and even begin to affect the local climate patterns. Choose statements from the box below and add them to Figure B.

Less rainfall	**Eroded soil may wash into rivers**	**Less water vapour in the air**
Rain hits ground with more force	**Less cloud formation**	

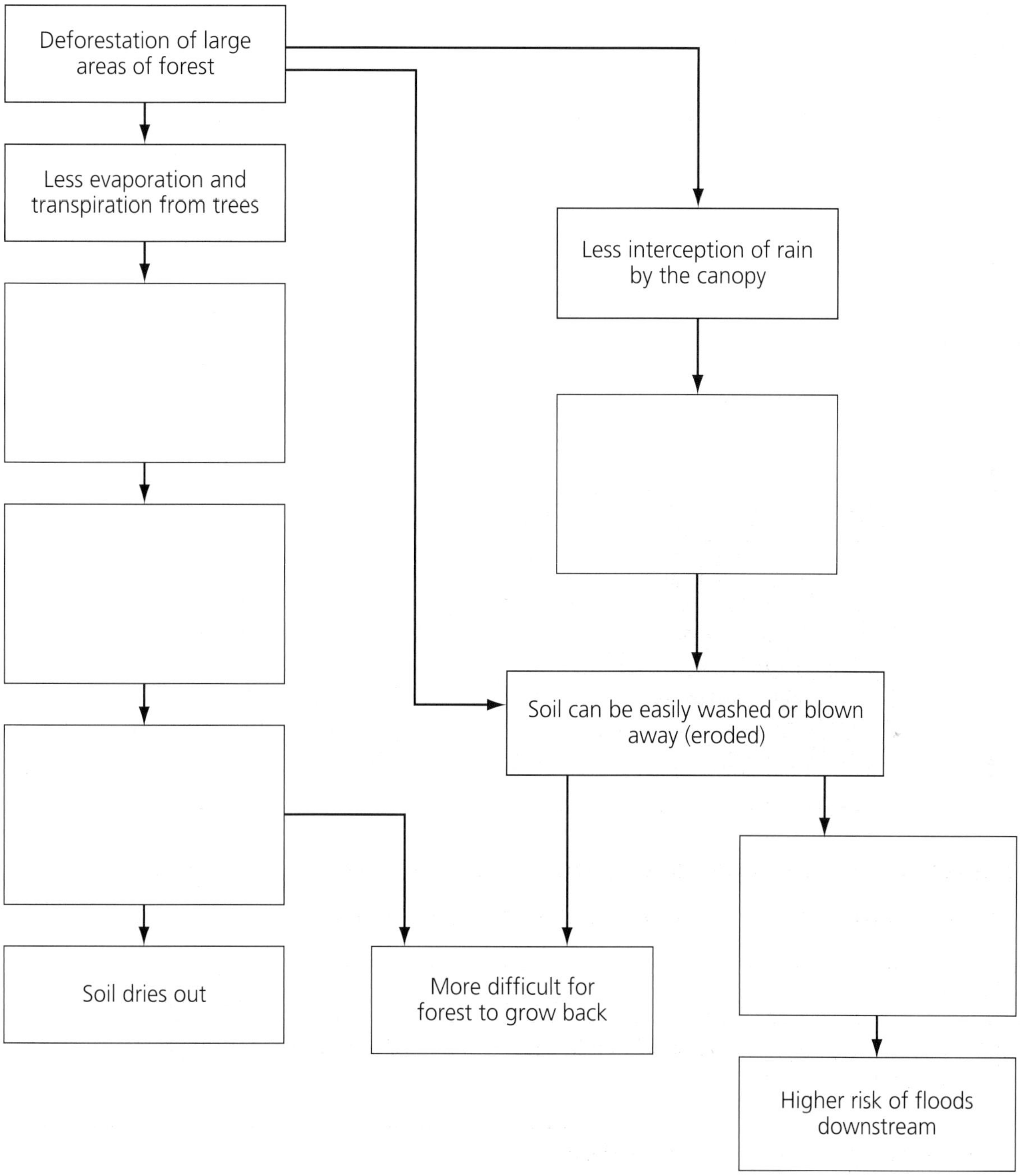

Figure B How deforestation may affect the water cycle

Review what you have learned

I can:

- explain why deforestation leads to soil erosion ☐
- describe how deforestation can affect local climate. ☐

Managing leisure use at Ynyslas

See Option Topics Book pages 58–59

In these activities you will:

- describe the location of Ynyslas
- describe how visitors have damaged the ecosystem here
- say how successful the management strategies here have been.

Case study: Ynyslas National Nature Reserve

1 Use Figure A to describe the location of Ynyslas National Nature Reserve by underlining the correct answers.

Ynyslas is located in **west /east / south** Wales. It is:

- north of the town of **Conwy / Aberystwyth / Wrexham**
- **100 / 120 / 150 km** from Cardiff
- on the coast next to **Cardigan Bay / Bristol Channel**
- located south of **Pembrokeshire / Snowdonia / Brecon Beacons** national park.

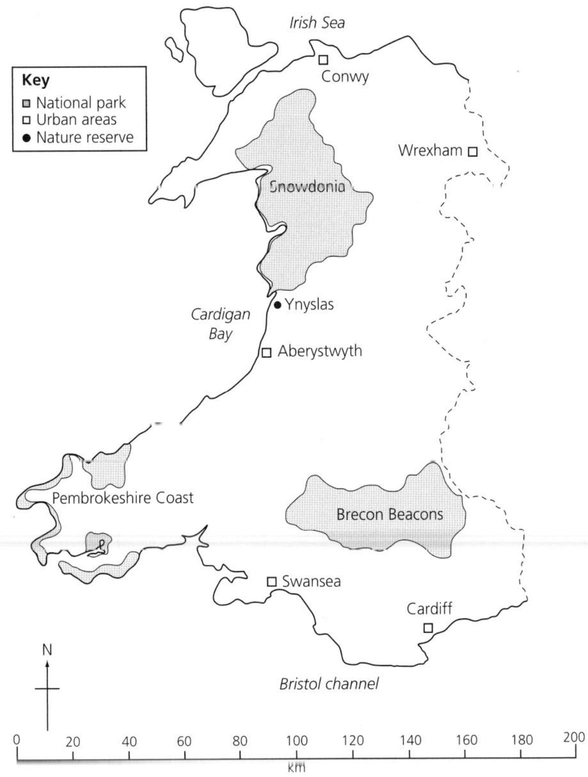

Figure A The location of Ynyslas National Nature Reserve

2 Human activities can cause erosion of sand dunes. Read the labels on Figure 28 on pages 58–59 and complete the diagram below.

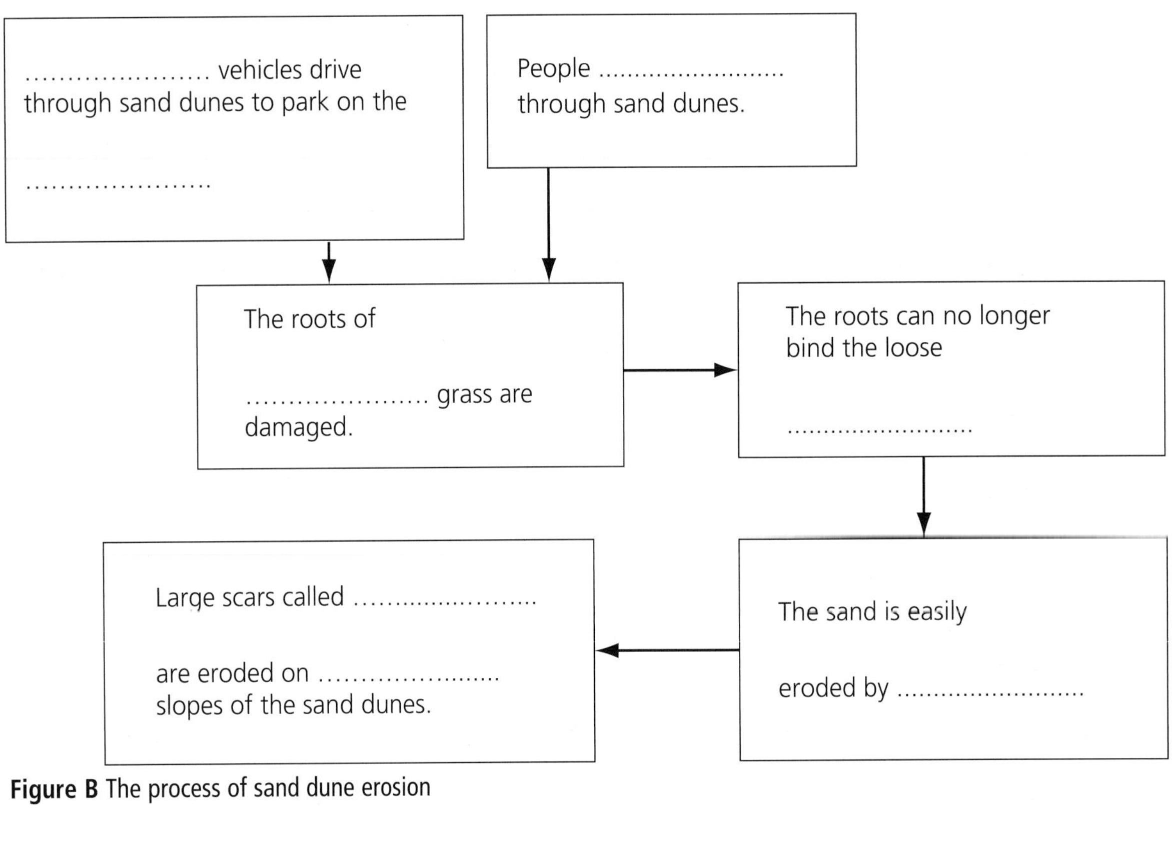

Figure B The process of sand dune erosion

3 Wardens at Ynyslas have introduced a number of strategies to try to manage the damage caused by human activity at the National Nature Reserve.

a) Read the information in Figure 28 on page 58 and 59 about what the wardens have done.

b) Complete the table below to evaluate the strategies. One has been done for you.

Damage or threat	How it is being managed	Evaluation of strategy
1. Driving vehicles through the dunes causes erosion	Put posts into the beach	Successful because cars cannot be driven onto the beach so plants grow and prevent further wind erosion
2. Visitors trample through the sand dunes		
3. Litter		
4. Visitors disturb the songbirds nesting on the ground		
5. Dog fouling		

✎ **4 a)** Draw a sketch map of the Ynyslas National Nature Reserve. Use Figure 28 on pages 58–59 to help you.

b) Add numbers from the table above to show where there are problems.

c) Add labels to show how these problems are being managed.

Review what you have learned

I can:

- describe a location using a compass point and scale line ☐
- describe three different ways that visitors cause damage to the dunes at Ynyslas ☐
- describe three management strategies and say how successful they have been. ☐

Tourism

What factors affect the nature of tourism in Cancun?

See Option Topics Book pages 67–69

In these activities you will:

- use compass points and a scale line to describe a location
- consider why Cancun is a good location for a large tourist resort
- practise the technique of drawing a climate graph.

In the 1970s, Cancun in Mexico was a tiny fishing village. Planners decided it was the perfect location for a large tourist resort. Why?

1 What makes Cancun a good location for a large holiday resort? Match the following statements to the factors shown on the spider diagram below. Some statements can be used twice. Two have been done for you.

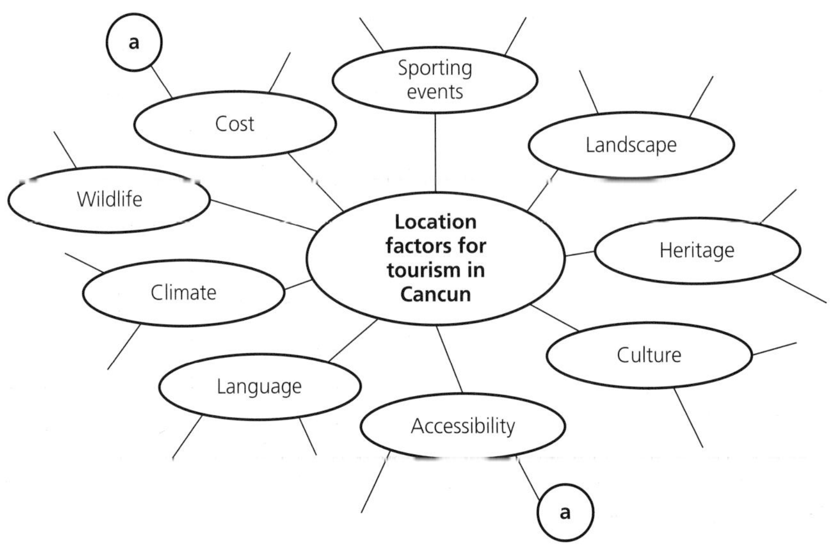

Figure A Location factors for Cancun

a It is easy to get there cheaply by air from the USA.	**f** There are plenty of places to dive on the reef.
b People can surf and water ski.	**g** Guides can take visitors turtle watching.
c It is warm and sunny for most of the year.	**h** There are miles of sandy beaches.
d People can go bird-watching in the rainforest.	**i** The resort has an airport nearby.
e People can visit Mayan ruins (like Figure 5) and museums.	**j** The sea is warm which encourages water sports.

2 Use Figure 4 on page 68 to complete the following sentences.

Cancun is in the state of **Quintana Roo** / **Yucatan** / **Belize**. It is on the coastline of the .... Sea. It is .... km north-east of the resort of Playa del Carmen. Chichen Itza is 220 km to the **north-west** / **north-east** / **west-south-west**.

3 Describe the location of Tulum. Use your answer to activity 1 to help structure your answer.

4 Complete the climate graph for Cancun, shown in Figure B.

a) Add three bars for the rainfall in April, May and June, based on the table below.

	Apr	May	Jun
Rainfall (mm)	30	80	170

b) Plot points, in the middle of each month, for the temperature. The first two have been done for you.

c) Join the temperature plots with a red line.

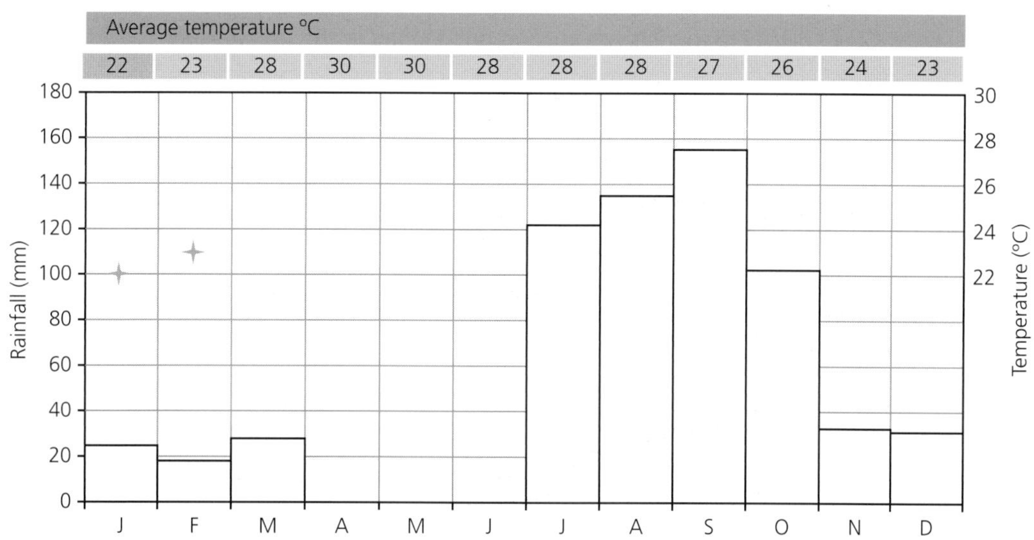

Figure B Climate graph for Cancun

5 Use Figures A and B and Figure 4 on page 68 to write an article on Cancun for a holiday brochure. Make sure you mention:

- climate
- location
- beach
- warm water.
- water sports

Review what you have learned

I can:

- use compass points and a scale line to describe a location ☐
- explain why the tourist resort was developed at Cancun giving three location factors ☐
- draw a climate graph with confidence. ☐

What are the advantages and disadvantages of mass tourism?

See Option Topics Book pages 78–80

In these activities you will:

- show you understand key geographical terms to do with tourism
- describe the advantages and disadvantages of mass tourism
- consider whether mass tourism in Cancun is sustainable.

1 Match the following key words to their definitions in the table below.

a) Direct jobs created by tourism
b) Indirect jobs created by tourism
c) Positive multipliers
d) Enclave tourism
e) Mass tourism

Key word	Definition
	Where the benefits of a new industry spread through the local economy
	Jobs that already exist but which benefit from tourism such as taxi drivers
	Where large numbers of tourists visit the same resort
	Where tourists are encouraged to stay in the hotel/resort
	Jobs in tourism industries such as hotel workers

2 Read pages 78–79. Describe the effects of each on local people or local businesses. One has been done for you.

a) There are a lot of jobs in hotels and bars so … *a lot of people have moved to Cancun to find work.*

b) Most jobs are unskilled and badly paid so …

c) Tourists stay in their hotels and rarely visit local bars and shops so …

d) Most airlines that bring tourists to Cancun are owned by American or European companies so …

Is mass tourism sustainable in Cancun?

Many people think that mass tourism in Cancun has been unsustainable. Read the list of statements below. They are based on pages 78–79.

1. Wages in bars and hotels are low.	9. Informal 'shacks' are small and overcrowded.
2. Informal workers are hassled by police.	10. Hotels use a lot of fresh water.
3. Drug and alcohol misuse are becoming a problem.	11. Informal 'shacks' are badly built.
4. The reef is damaged by boats and divers.	12. People in shanty towns have to share toilets.
5. Working hours are long.	13. Local people are not allowed on the beach.
6. The coral reef has been damaged by sewage.	14. The lagoon is polluted by oil from boats.
7. Working conditions are stressful.	15. Hotels import expensive food.
8. Rents for formal housing are very high.	16. Air conditioning in hotels uses a lot of energy.

3 Study Figure 33 on page 80 of the Option Topics book and the list on page 78 of this workbook. Choose at least eight statements from the list and add them to the copy of Figure 33 below to show why tourism may be **unsustainable** in Cancun. One has been done for you.

Local people need to benefit. 13	The environment should not be damaged so much it cannot recover.
The growth of tourism should not create problems for future generations of people.	The growth of tourism should not create so many problems that people stop coming.

4 Choose one statement from each box above. Write at least three suggestions about what needs to be done in Cancun to make tourism **more sustainable**.

5 Tourism in Cancun has damaged the coral reef. What do you think should be done to reduce this damage?

a) Read the nine possible solutions below and add them to the diamond shape. Put the numbers of the solutions that you think would work best at the top.

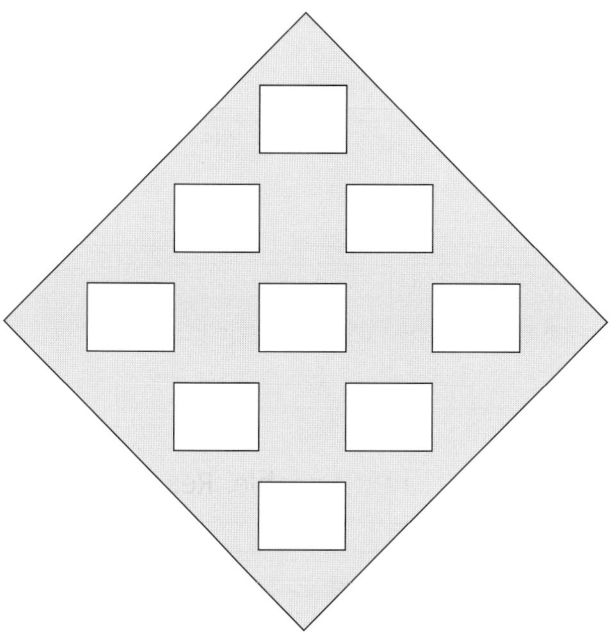

1. Allow divers to use only one section of reef.
2. Spread divers over a much wider section of reef.
3. Ban all divers from the reef for five years.
4. Educate divers about the reef system.
5. Improve sewage treatment in Cancun.
6. Make rainforest tours cheaper.
7. Make diving much more expensive.
8. Spend money on scientific research on the reef.
9. Create a new dive site by sinking an old ship.

b) Explain why you think your chosen solutions are best.

Review what you have learned

I can:

- use key terms with confidence ☐
- describe one advantage and two disadvantages of mass tourism ☐
- give three reasons why mass tourism may not be sustainable for Cancun. ☐

The future of air travel

See Option Topics Book pages 74–75

In these activities you will see that:

- people have conflicting views over whether the UK needs a new airport
- when planners make decisions, they consider different points of view.

1 Study Figure 22 on page 75 and the statements in the boxes below.

a) Decide whether each statement is **for** or **against** the expansion of airports.

b) Decide whether the statement is making a **social** argument, an **economic** argument or an **environmental** argument.

	For (✓) or against (✗)	Social Economic Environmental
A Expanding airports will create new jobs in construction. There will also be new jobs at the new airports.		
B Some homes will be bulldozed to make space for new runways. These home owners will have to move from their community.		
C The expansion of airports is good for the UK economy because it means more foreign visitors.		
D Carbon dioxide released from aircraft is contributing to climate change.		
E Noise from aircraft landing and taking off is stressful for home owners who live under the flight path. It can reduce house prices.		
F Expanding UK airports would mean a loss of farmland, which could mean more reliance on importing food in the future.		
G Technology is being developed to make aircraft cleaner and more efficient. This will reduce their impact on climate change.		
H Passengers could be charged higher taxes when using air travel to try to reduce the number of people flying.		
I Some air companies encourage passengers to pay a charge which is used to plant trees. This helps to offset the impact of aircraft on the atmosphere.		

2 Study Figure 21 on page 74. Name one group who might be:

a) in favour of a new runway at Heathrow: ....

Explain why they have this view. ....

....

b) against a new runway at Heathrow: ....

Explain why they have this view. ....

....

c) against a new airport at Cliffe: ....

Explain why they have this view. ....

....

3 How can the problem be solved? Use information from pages 74 and 75 to complete the following table. You could also research the issue on the internet.

A Build a new runway at Heathrow	Positive impacts of this decision	
	Negative impacts of this decision	
B Build a brand-new airport at Cliffe in the Thames Estuary	Positive impacts of this decision	
	Negative impacts of this decision	

4 Decide which of these two alternatives you think is the best option for the UK. In your answer you should include:

- which option you think is best
- why you think it is the best option (give at least two reasons)
- how you think the negative impacts of your decision could be reduced.

Review what you have learned

I can:

- name two different groups of people who have strong opinions about this issue ☐
- summarise the main argument used by each of these groups. ☐

Is tourism good or bad for Iceland?

See Option Topics Book pages 76–77, 89

In these activities you will:

- consider some problems created by tourism in rural parts of Iceland
- describe the benefits of tourism for the economy of Iceland.

1 Match the following key words to their definitions in the table below:

a) wilderness **b)** honeypot sites **c)** positive multiplier effect **d)** rural diversification

Key word	Definition
	A chain of events, triggered by investment or the creation of new jobs, that bring wider benefits to the local economy
	Landscapes that are unspoilt by people – these landscapes are largely uninhabited and not farmed
	Places or visitor attractions that are so popular that they become congested with tourists, which may lead to damage to the environment
	The changing nature of jobs in the countryside, e.g. from farming to tourism

2 Read pages 76–77.

a) Use information from these pages and Figure 26 to complete the matching exercise below.

People from Reykjavik buy second homes in rural areas that they only use in the summer so …	… some landscapes in the Golden Circle have become honeypot sites which are damaged by the number of visitors.
Volcanic soils are very loose so …	… where plants have been damaged by visitors the landscape can take many years to recover.
Plants grow very slowly in the cool climate and the summer growing season is very short so …	… rural communities need to make most of their income during the summer months.
Tourists like to see open spaces that are wild and unspoiled but close to main roads so …	… they are easily damaged by off-road driving.
Iceland has very few trees so …	… there may not be enough local people to keep rural schools or shops open in the winter.
More visitors come to Iceland during the summer than the winter months so …	… there is little to stop the wind from eroding the soil if plants become trampled by people or vehicles.

b) Identify **one** social and **one** environmental problem caused by tourism in rural Iceland:

Social ....

Environmental ....

3 Tourism and leisure activities create a lot of benefits in the economy of the rural area.

a) Study page 76 and Figure A below, which shows the effect of a **positive multiplier** on the area around Gullfoss.

b) Complete the blank spaces in Figure A by adding words from Boxes 1 and 2 below.

c) Use **two** colours to shade the boxes in Figure A to show the cause and effects of tourism.

Box 1

roads, schools and hospitals
car hire companies and coach companies
shops and restaurants
banks and building societies
waiters, cleaners and hotel receptionists

Box 2

more
less
the same

You will need to use words from this box more than once.

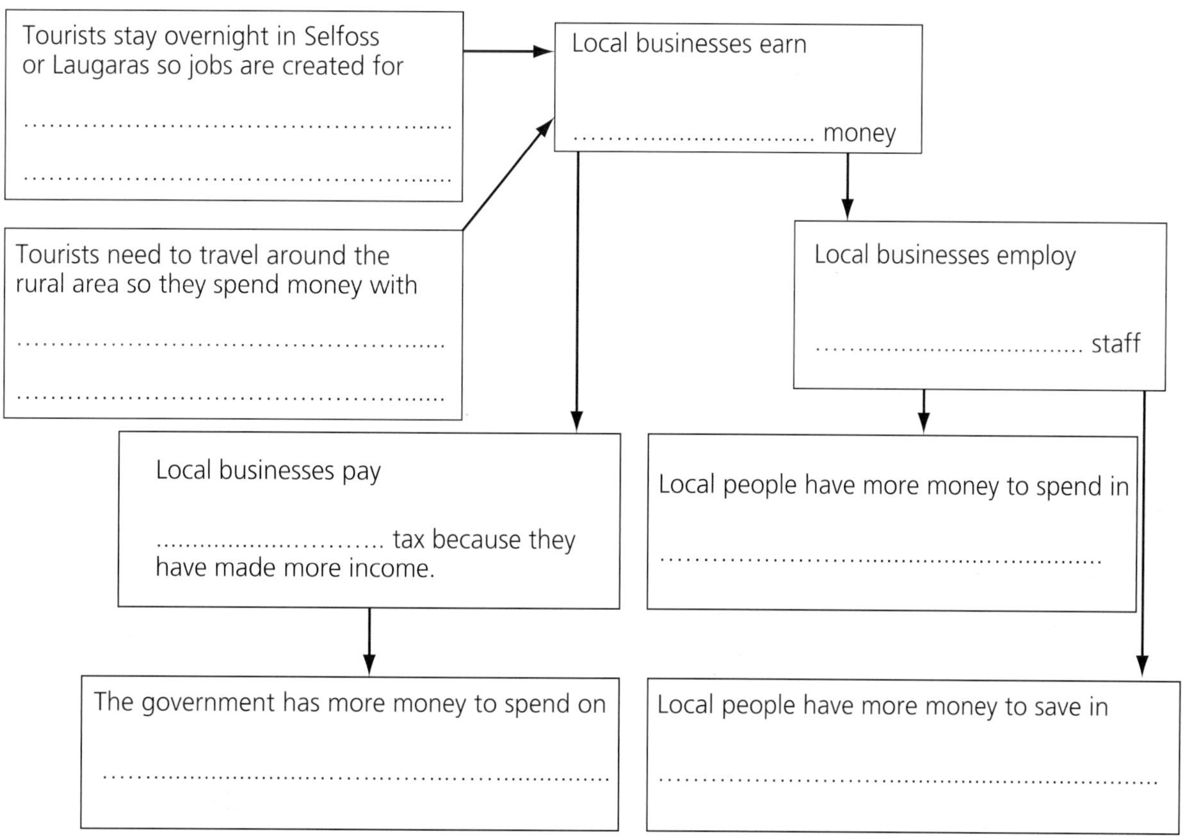

Figure A The positive multiplier effect in rural Iceland

Review what you have learned

I can:

- describe one social problem created by tourism in rural parts of Iceland ☐
- describe one environmental problem created by tourism in Iceland ☐
- describe three benefits of tourism for the economy of Iceland. ☐

Retail and Urban Change

Urban regeneration in Barcelona

See Option Topics Book pages 94–95

In these activities you will:

- develop a case study of urban change in Barcelona
- examine how El Raval was improved through a process of gentrification.

1 Study Figure 12 on page 94.

a) Underline the correct answers to complete this description of the location of Barcelona.

Barcelona is in **Catalonia** / **Andorra** in Spain. It is located about **50** / **80** / **110**

kilometres **north-east** / **south-west** of Tarragona.

b) To describe a location you need to use the compass point and scale line. Use your answers to activity 1 a) as a model to describe the location of two more cities in relation to Barcelona:

i) Lerida **ii)** Figueras.

2 Barcelona grew very quickly between 1930 and today. This caused a number of problems. Read the text on page 94 carefully and then match the following words to the correct sentences in the table below.

a) mass clearance of homes **b)** pressure on transport systems

c) regeneration of huge brownfield sites **d)** suburban sprawl

Key word	Definition	
	Barcelona grew from just 60 km^2 to 500 km^2 so that many people live many miles from the city centre.	Problems
	Roads and railways struggled to cope with the increasing population. Traffic jams became a problem.	↓
	Many old flats were demolished in El Raval so that room could be made for new housing, museums and restaurants.	↓
	Old industrial areas near to the port were replaced with sporting venues and hotels.	Solutions

3 Discuss Figure 15 on page 94. What do you think about these modern planning ideas? Choose one of them and use the writing frame that follows to explain the advantages of your choice.

I think the most important key principle is …

This principle has advantages for local people because …

It will benefit the urban environment because …

In the long term the effect of this would be …

Gentrification in Barcelona

Planners in Barcelona have tried to improve the overcrowded areas of the city such as El Raval. El Raval is now becoming a more popular neighbourhood for young families to live in. This process is known as **gentrification**. Study the information on pages 94 and 95 before answering the following questions.

4 Study the statements in the boxes below. Sort the statements into two types:

- **problems** that planners needed to solve to improve the city
- **solutions** that the planners have used.

Colour the problems in red and the solutions in green.

Planners have demolished some of the old apartment blocks in El Raval so the population density is now lower.	Public transport has been improved. It is harder for car owners to drive around the city so it is safer to walk around.
Drug dealing and prostitution were common in El Raval so tourists were warned not to visit the area.	The run-down, industrial areas around the port of Barcelona put people off visiting the city.
New museums and art galleries have been opened so people have more leisure opportunities.	New parks and gardens have been created so there are places to relax and safe places for children to play.
When people made a noise late at night (for example, shouting or using motorbikes) the residents of El Raval could not sleep.	The city authorities started to fine people on the spot for making too much noise so there is a more peaceful environment.
People are fined for vandalism such as graffiti so that the local residents have pride in the area and it attracts visitors.	The local authority listened to the views of local residents and provided them with what they wanted.
The streets of El Raval were narrow and congested. It was difficult for people to get to work and for local businesses to run efficiently.	In 1980, 4 per cent of people in Barcelona lived in homes that were unfit to live in and many more people lived in overcrowded conditions.

✎ **5** Study the statements in the boxes above. Imagine you are a planner.

a) Rank the problems starting with the one that needs solving first. Explain why you would tackle this problem.

b) Choose the solution to this problem. Say how successful you think it has been.

Review what you have learned

I can:

- describe three problems faced by planners in El Raval ☐
- describe how these three problems were solved. ☐

Retail change

See Option Topics Book pages 96–99

In these activities you will:

- describe patterns of retailing in a typical UK town
- consider how planners can protect shops in town centres.

1 Study Figure 19 on page 96 and Figure A below. It shows a typical pattern of retail provision in UK cities. In each Fact box write down **one** important feature and **one** recent change. Fact box 3 has been done for you.

Key

CBD : Central Business District
(hatched symbol) : Urban area
① - ⑤ : See fact boxes
(arrow) : Main roads

Fact box 1

Fact box 2

Fact box 3

- Feature: There are many newsagents/off-licences/take-away outlets.
- Change: Many of the original shops (butcher/greengrocer) have closed through competition from supermarkets.

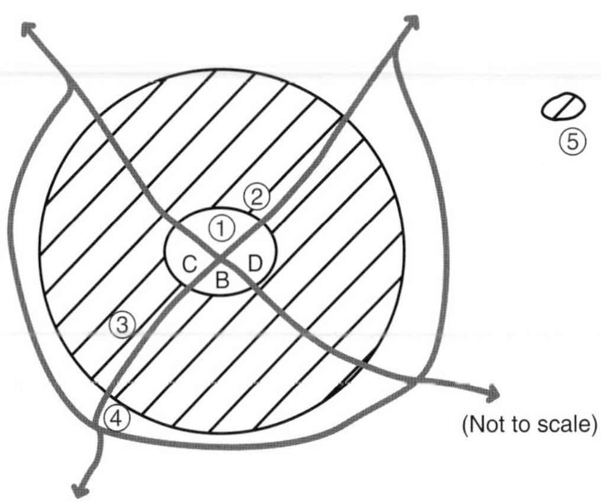

Fact box 4

Fact box 5

Figure A The typical pattern of retail provision in UK cities

2 Study pages 98 and 99. These pages describe how planners in Shrewsbury have tried to retain many of its older independent shops.

a) Study Figure 23.

i) Underline the correct answers to describe the location of Sundorne retail park.

Sundorne retail park is **2** / **3** / **4** km **north-east** / **south-west** of Shrewsbury town centre.

✎ **ii)** Describe the location of Meole Brace retail park.

✎ **iii)** Use Figure 24 to give two reasons why some people prefer to shop in Shrewsbury's retail parks rather than the town centre.

Reason one: ....

Reason two: ....

3 Discuss how Shrewsbury can attract more shoppers to the town centre.

a) Put the following solutions into rank order. Put the solution you think will have the biggest impact as number 1 and write the rank in the boxes.

- ☐ Improve town centre parking
- ☐ Improve public open space for shoppers
- ☐ Increase Sunday trading
- ☐ Encourage continental café style culture
- ☐ Reduce rates for empty shops

✎ **b)** Explain how the top three solutions will help retailers and shoppers in Shrewsbury.

4 Shrewsbury is doing its best to support businesses in the town centre. One example is a scheme called 'All the Little Shops'. Study Figure 25 on page 99.

a) Underline the correct answers in the passage below.

Independent retailers / **chain stores** in Shrewsbury have been given funding to create their own webpage where up to **30** / **50** products can be displayed and traded **at a small charge** / **free**.

✎ **b)** Work in pairs. Discuss Figure 25 before answering the following questions:

i) What does this source tell me about the Little Shops scheme?

ii) What guesses can I make about the scheme? What can I infer?

iii) What doesn't this source tell me?

iv) What other questions do I need to ask about the Little Shops scheme?

Review what you have learned

I can:

- describe typical patterns of retailing ☐
- describe three strategies that aim to help independent retailers. ☐

How has the internet changed retailing?

See Option Topics Book pages 100–101

In these activities you will:

- recognise the changes brought by internet sales
- practise using an OS map to describe location factors.

1 Study page 100. Complete the boxes in Figure A below to explain why many music stores have shut down on the high street. One box has been done for you.

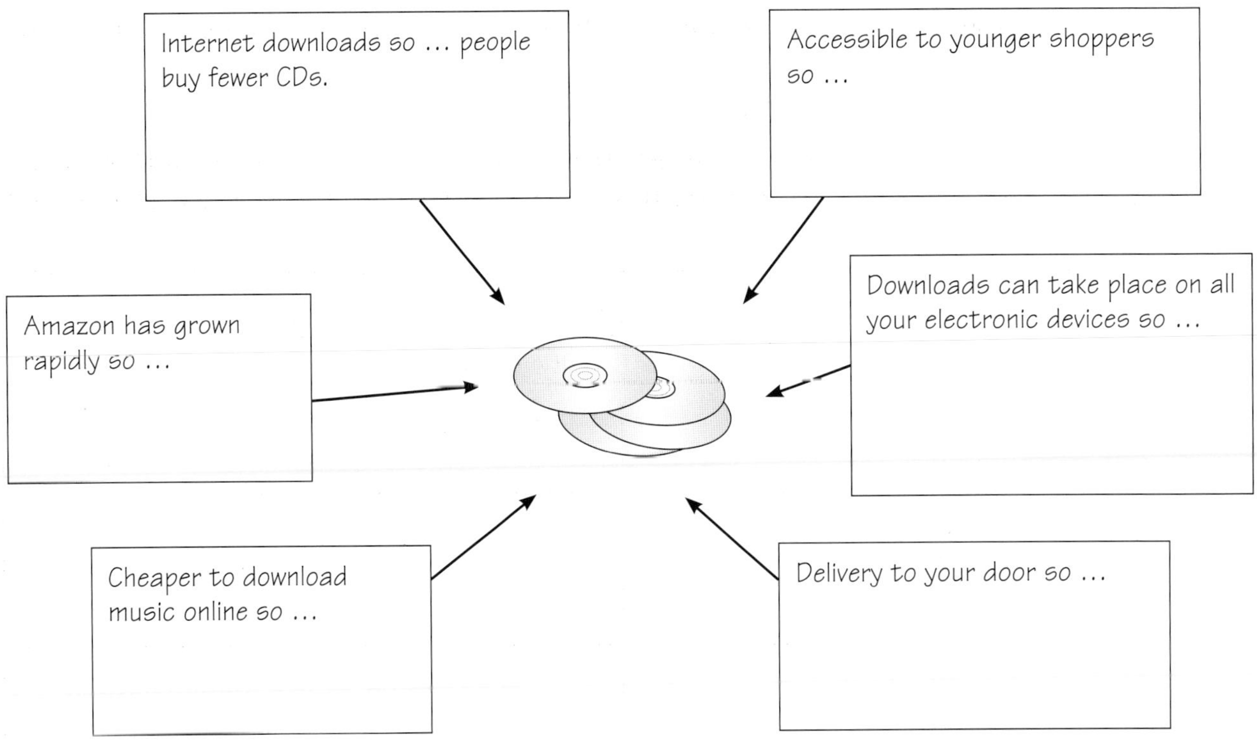

Figure A Why are high street music stores closing down?

2 Study Figure 28 on page 100. It shows the rise of UK internet shoppers. Underline the correct words in the passage below.

Between 2001 and 2006 the number of people shopping on the internet **fell** / **rose**. People using online banking **doubled** / **trebled**. People shopping online saw the **largest** / **smallest** increase from **11** / **13** million to **27** / **29** million people. Telephone banking saw the **largest** / **smallest** increase. Purchases per online shopper saw a **fall** / **rise** from **8** / **10** million to **23** / **27** million purchases.

✎ **3** Study Figure 29. Give three reasons why internet shopping is beneficial to the elderly group of people.

4 Amazon is a leading online company. It has a large warehouse in Swansea. Study Figures 30 and 31 on page 101. Complete the sentences opposite. Use the words from the word box.

B4290	**A483**	**expand**	**7193**	**accessibility**	**roundabout**	**built-up**
B1680	**steep**	**A683**	**7293**	**open**	**flat**	**motorway**

Amazon has opened its warehouse in grid reference .... South of the warehouse is the .... road which links to a .... and the .... road. This road links to the M4 .... These links give the warehouse good .... The land around the warehouse is .... which means it has been easy to build upon. There is plenty of .... space around the warehouse which means there is room to .... in the future.

5 The opening of the Amazon warehouse has received lots of positive and negative comments. Study page 101. Sort some positive and negative into the table below. One example has been done for you.

Positives	Negatives
	Many jobs are low paid.

✎ **6** Review your learning so far. Comment upon whether you think that the rise of internet shopping is good for the UK or not. Use the writing frame that follows.

The internet has a number of benefits for shoppers. I think the most important is … Another benefit for shoppers is …

Some people who work in the retail industry have also benefited from the internet. One example is …

They have benefited because …

However, internet shopping has created some problems for retailers such as …

Furthermore, some planners are worried about the impact of the internet on high street shops because …

Review what you have learned

I can:

- identify the positive and negative impacts of internet shopping ☐

How does retail affect people and the environment?

See Option Topics Book pages 106–109

In these activities you will:

- identify how shopping in the UK can affect people in other countries
- weigh up the arguments for and against a planning proposal.

1 Study Figure 42 on page 106 and Figure 45 on page 107. Figure 42 compares the average cost of clothes. Write whether the statements are true or false.

Statements	True or False?
It is more expensive to buy clothes from the average high street store.	
Matalan's clothes are cheaper than Tesco's.	
Primark produces the cheapest clothing compared to all other producers.	
Primark produces clothes at least a third cheaper than the high street store.	
Asda/George produces clothes half the price of Tesco's.	
Despite the recession Primark has seen an increase in sales.	
In 2010 Primark was planning to close stores in the UK and Europe.	

2 Study Figure 43 on page 107. The girl has mixed views about buying cheap clothes made in Asia. Use this information to help you complete the following sentences which explain her difficult choice. You could use some of the words in the box in your answer.

poverty	low wages	sweatshop	long hours	child labour

Perhaps I should buy these clothes because ....

....

On the other hand, perhaps I should avoid buying these clothes because ....

....

....

3 Study the information on page 106 about the fashion trade and ethics. Work in pairs. Discuss Figure 40 before answering the following questions:

a) What does this source tell me about Primark?

b) What guesses can I make? What can I infer?

c) What doesn't this source tell me?

d) What other questions do I need to ask about Primark?

4 Study Figures 47 and 48 on page 108. They give us information about what household waste is made up of and where it goes. Fill in the gaps by choosing words from the word box below.

fabric	**26**	**glass**	**recycled**	**paper/card**	**incinerated**
12	**metal**	**20**	**bio-treated**	**28**	**landfill**

Household waste is mainly made of .... at 33% followed

by meat/plant waste at ....%. Plastic contributes ....% with

.... and .... both at 11%. Most of the UK

waste goes to .... sites with only ....% recycled.

5 Study page 109. It gives details about a proposal to build a waste incinerator at Ffos-y-fran in Wales.

a) Complete the following sentences which give facts about the case study.

The incinerator would create .... jobs. The incinerator was

expected to generate .... of electricity which is enough for

.... homes.

b) A number of groups of people had a view about whether this proposal should go ahead. Explain why each of the following groups held their view:

Councillor Jeff Edwards supported the proposal because

....

....

Friends of the Earth campaigner, Haf Elgar, was against the proposal because

....

....

✎ **c)** In October 2011 Covanta withdrew its application to build the incinerator – there was too much opposition from local residents. Do you think the company should have gone ahead with the scheme or not? Use the writing frame below.

I would support/object to the scheme. The benefits of my decision for local people would be … My decision would be good for the environment because …

Review what you have learned

I can:

- describe how shoppers' decisions in the UK can affect people in other countries ☐
- identify fact and opinion when considering arguments for and against a planning proposal. ☐

Economic Change and Wales

Patterns of work and employment

See Option Topics Book pages 111–114, 119

In these activities you will:

- show you know the difference between different employment sectors
- describe the distribution pattern of some types of work in Wales.

1 Read page 112. Match each key word to the correct definition below.

a) primary sector **b)** secondary sector **c)** tertiary sector

d) public sector **e)** private sector

Key word	Definition
	People working in this sector provide a service.
	People working in this sector are employed by national or local government.
	People working in this sector produce raw materials or grow food.
	People working in this sector are employed in a manufacturing or processing industry.
	People working in this sector are employed by a company or are self-employed.

2 Sort the following jobs into sectors. Write **P** for primary, **S** for secondary or **T** for tertiary in the box beside each job. The first one has been done for you.

Farmer	P	Forestry worker	
Teacher		Self-employed furniture maker	
Baker		Lorry driver delivering food to a supermarket	
TV / film director		Lawyer	
Bricklayer on a building site		Software programmer writing computer games	
Cleaner in a hotel		Worker in a factory that tins vegetables	

3 Study Figures 6 and 7 on page 113 before answering the following questions. You may need to refer to the place names on Figure 16 on page 119 or use an atlas or Google Earth to help you.

a) Where are the following towns and cities located? Are they in East or West Wales? Put a tick in the correct column (✓).

Town/city	East	West	Town/city	East	West	Town/city	East	West
Cardiff			Mold			Wrexham		
Newport			Monmouth			Oswestry		
Swansea			Conwy			Bridgend		
Carmarthen								

b) Underline the correct answers in the following statements:

Over 10 per cent of people in Powys and **Gwynedd** / **Ceredigion** / **Cardiff** are employed in farming. Farming is also an important employer in Carmarthenshire and Pembrokeshire, both of which have **1–2.9** / **3–4.9** / **5–9.9** per cent employed in farming. Farming is less common in Valleys communities in South Wales such as Rhonda Cynon Taff where **0–0.9** / **1–2.9** / **3–4.9** per cent are employed in farming. Overall, West Wales and the valleys have about **60** / **70** / **80** per cent of all of Wales' farming. East Wales has about **45** / **50** / **55** per cent of all of Wales' jobs in finance.

4 Study Figure 8 and Figure 9 on page 114. You may need to refer to the place names on Figure 16 on page 119 or use an atlas or Google Earth to help you.

a) Underline the correct answers in the following statements:

The region with the largest percentage employed in manufacturing is **Flintshire** / **Denbighshire** / **Wrexham**. The parts of West Wales that have most employment in manufacturing tend to be in the south and Valleys communities such as **Carmarthenshire** / **Bridgend** and **Torfaen** / **Cardiff**. The region with the smallest percentage employed in manufacturing is in **Denbighshire** / **Conwy** / **Gwynedd**.

✎ **b)** Describe the distribution of aerospace industries. Use Figure 16 on page 119 to help you. Also use appropriate words from the box below.

regular	**even**	**uneven**	**irregular**	**clustered**

5 Study each of the following lists of words.

a) In each list, circle the odd one out.

					Common theme
1	teacher	secondary	manufacturing	factory	
2	farming	forestry	banking	mining	
3	Land Rover	Tesco	Sony	Nokia	

b) For each list, state what the three remaining words have in common. Put your answer in the final column of the table.

Review what you have learned

I can:

- give three examples each of primary, secondary and tertiary jobs ☐
- describe patterns of employment in farming and manufacturing. ☐

How and why is employment changing in Wales?

See Option Topics Book pages 112, 116–119

In these activities you will:

- practise drawing a bar graph that has positive and negative values
- describe ways that employment in Wales is changing.

1 Study Figure 4 on page 112. It shows how the proportion of people employed in the public and private sectors in Wales has been changing. Use information from this graph to complete the following sentences by underlining the correct answers.

Between 1999 and 2002 jobs grew more quickly in the **private** / **public** sector. The number of people employed in the public sector reached a peak in **2002** / **2003** / **2004**. Since that date the number of people employed in the private sector has tended to grow **faster** / **more slowly** / **at the same rate** as jobs in the public sector.

2 Read pages 116–118. Match each key word to the correct definition below.

a) mechanisation
b) multi-national companies
c) de-industrialisation
d) positive multipliers

Key word	Definition
	Where the benefits of a new industry spread through the local economy
	The use of machines to replace human labour
	The decline of traditional manufacturing jobs
	Businesses that have factories and offices in a number of different countries

3 What factors attracted Amazon to locate their warehouse near Swansea? Read page 116 and study Figure 12 on page 117. Join the heads and tails below to help explain the decision.

The M4 motorway is only 2 km away so …	… Amazon could recruit local people to work in the warehouse.
The land where the warehouse was built was a brownfield site so …	… services such as gas, electricity and water were already on site.
Swansea, with a population of 230,000, is only 4 km away so …	… deliveries can quickly be made to customers in South Wales and South England.
The site was large and flat so …	… the company could recruit managers who would be prepared to move into the area.
The coastline of the Gower is nearby so …	… it was easy to build a very large warehouse with space for delivery lorries to park.

4 Study Figure 10 on page 116. It shows how employment changed between 2001 and 2006.

a) Complete the values on the horizontal axis in Figure A. Give this axis a label.

b) Use the data to complete Figure A. Two bars have been drawn for you.

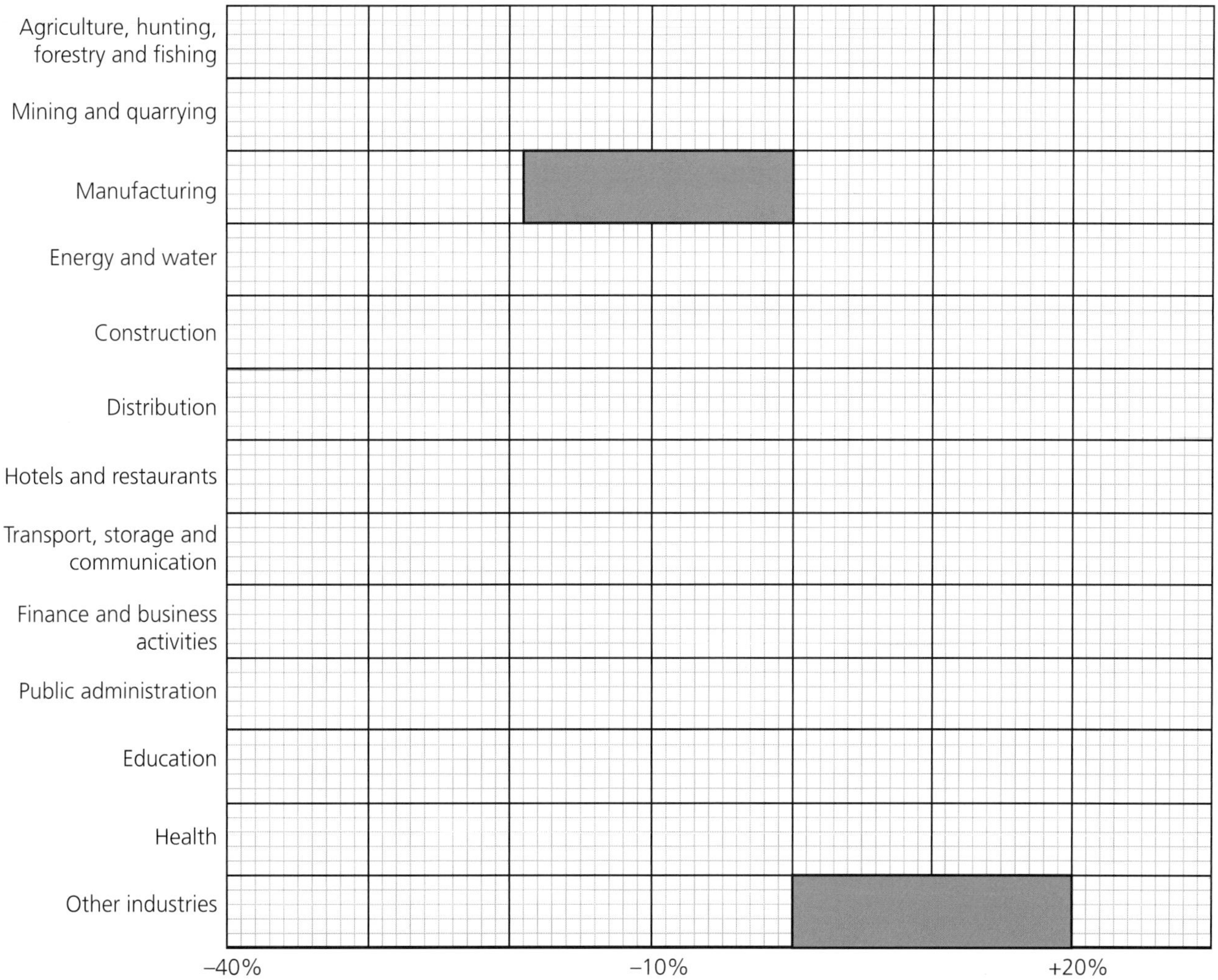

Figure A Changes in employment in Wales (2001–2006)

c) Read page 118 and use your completed graph. Summarise what you have learned by completing the sentences below. Use words from the box to help you.

service **primary** **retraining** **public** **'heavy' manufacturing** **de-industrialisation**

There has been a decline in .... in Valleys towns such as Merthyr Tydfil. Between 2001 and 2006 employment in manufacturing declined by **18** / **20** / **28** per cent. However, new jobs have been created, mainly in the .... sector such as retailing. .... has created problems for some local people and unemployment **rose sharply** / **fell** in Wales between 2008 and 2009. People need .... before they are ready for new jobs.

Review what you have learned

I can:

- present data on a graph that has positive and negative values ☐
- describe ways that employment in Wales is changing. ☐

Changes to employment in rural Wales

See Option Topics Book pages 119–124

In these activities you will:

- examine some of the issues facing communities in rural Wales.

1 Study Figure 17 on page 120. You may need to refer to the place names in Figure 16 on page 119 or use an atlas or Google Earth to help you.

a) Underline the correct answers in the following statements:

Large parts of Wales are difficult to farm. The EU classifies about **40** / **60** / **80** per cent of Wales as Severely Disadvantaged Area (SDA). Nearly all of Powys and **Gwynedd** / **Pembrokeshire** are classified SDA.

The EU classifies a further **20** / **40** / **60** per cent of Wales as Disadvantaged Area (DA). Large parts of Anglesey and **Carmarthenshire** / **Vale of Glamorgan** are classified DA.

b) Study the list of diversification schemes in the table below. They are all mentioned in Figure 17. Add the missing locations to the second column. Some have been done for you.

	Type of diversification	Location
1	Growing oilseed rape for biofuel	
2	Food markets	Pwllheli
3	Native plants/trees are grown from seed	Rhiw
4	10 hectares of woodland planted	
5	Production of organic milk drinks	
6	Historic visitor attraction and tea room	Trapp
7	Fly fishing	
8	Paintball games	
9	Daffodils for medicine	
10	Miscanthus for biofuel	Penllyn Estate
11	Green waste for composting	
12	Rock climbing and abseiling	
13	Hedge trimming and dredging	Penybont
14	Growing and selling leeks	Queensferry

c) Work in pairs to sort the schemes into different types by placing the number of each scheme onto the Venn diagram in Figure A below. One has been done for you.

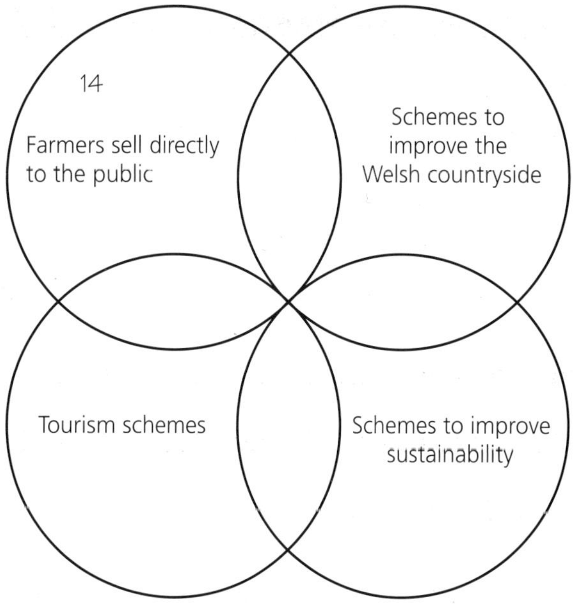

Figure A Types of farm diversification schemes in Wales

2 Read page 122 and study Figure 19 on page 123. Use information from these pages to join the heads and tails below.

Some young people do not want to work in the farming industry so …	… the price of rural housing increases.
There are fewer young people in the rural population so …	… they move away to get a job in an urban area.
Demand for local shops, schools and pubs falls so …	… some of these services may close.
Demand from second home owners increases so …	… more food has to be transported long distances, which increases its carbon footprint.
Local people in poorly paid jobs cannot afford to buy rural homes so …	… the result is an ageing population with less demand for schools and other services.
Because young people are reluctant to become farmers there is a fear that less food is grown locally so …	… they move to parts of Wales where housing is cheaper.

✎ **3** Discuss Figure 21 on page 124 before answering the following questions:

a) What does this source tell me about future jobs in Wales?

b) What guesses can I make about the future? What can I infer?

c) What doesn't this source tell me?

d) What other questions do I need to ask the Welsh Government about future jobs in Wales?

Review what you have learned

I can:

- describe three different types of farm diversification ☐
- describe three possible social impacts of rural change. ☐

Should we have tidal power in the Severn estuary?

See Option Topics Book pages 130–132

In these activities you will:

- practise your decision-making skills.

1 Study Figure 33 on pages 130–131.

a) Use information from this diagram to complete the columns marked * in Figure A. Some parts of the table have been done for you.

Option	* Amount of energy provided (per cent of UK electricity demand)	† Rank order	* Construction cost (£ billions)	† Rank order
1 Beachley Barrage				
2 Inner Barrage	1		1.4–1.8	
3 Onshore lagoon (Wales)				
4 Onshore lagoon (England)				
5 Middle Barrage				

Figure A Economic cost/benefits

b) Put the information in each column in rank order starting with the highest number. Put your values in the columns marked †.

✎ **c)** Use your table to suggest the best economic option. Explain your choice.

2 Read Figure 34 on page 132. Opinions about the benefits and impacts of tidal power in the Severn estuary vary. Read the statements below.

A 'The Severn estuary includes several large mud flats. These "intertidal habitats" are areas of mud that are exposed at low tide. They attract wading birds which feed on worms and shellfish living in the mud.'

B 'Newport and Avonmouth are important UK ports. Some of the barrage options could reduce ships' access to the ports and threaten jobs. The jobs of 7,500 people depend on the port at Avonmouth.'

C 'The Gwent Levels are a flat, low-lying coastal area in South Wales. This coastline needs protection from the threat of coastal flooding if sea levels rise. Some of the tidal power options would prevent coastal floods.'

D 'Tidal power is clean and renewable. It could reduce the UK's reliance on fossil fuels to generate electricity so we would reduce our CO_2 emissions.'

E 'Building the barrage will use a huge amount of concrete. Making concrete creates a lot of CO_2 emissions.'

a) Match the statements to the following groups of people. Do you think they are in favour of the tidal power schemes or not?

Group	Statement (A–E)	Likely to be in favour of a tidal power scheme or not?
Landowners in Caldicott, east of Newport		
Business leaders in Bristol		
Friends of the Earth Cymru		
Spokesperson at Slimbridge		
Climate change scientist		
Bristol Port Company		
The Welsh Government		

b) Study Figure 33 again.

i) Which option could make the biggest impact on climate change? ....

ii) Which option could have the worst effects on ports? ....

iii) Which two options would help protect the low-lying land of Gwent? ....

iv) Which two options would have no effect on major ports? ....

✎ **c)** Use Figure 33, as well as your answers above, to decide which is the best option. Use the writing frame below to structure your answer.

I would advise that the best option is …

My main reason for choosing this option is …

My option has the following economic benefits: …

Furthermore, it has these environmental benefits: …

It is better than the other options because …

In conclusion, I think my option is the most sustainable because …

Review what you have learned

I can:

- use rank order technique to help make a decision ☐
- identify economic and environmental impacts of tidal power. ☐

Notes